Selected Titles in This Series

(Continued in the back of this publication)

Mathematics of Fractals

Translations of
MATHEMATICAL MONOGRAPHS

Volume 167

Mathematics of Fractals

Masaya Yamaguti
Masayoshi Hata
Jun Kigami

Translated by
Kiki Hudson

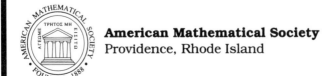

American Mathematical Society
Providence, Rhode Island

フラクタルの数理

(FURAKUTARU NO SŪRI)

(Mathematics of Fractals)

by Masaya Yamaguti, Masayoshi Hata, and Jun Kigami

Copyright © 1993 by Masaya Yamaguti, Masayoshi Hata, and Jun Kigami

Originally published in Japanese by Iwanami Shoten, Publishers, Tokyo, 1993

Translated from the Japanese by Kiki Hudson

1991 *Mathematics Subject Classification.* Primary 58F13; Secondary 70K50.

ABSTRACT. This book aims at providing a handy explanation of the notions behind the self-similar sets called "fractals" and "chaotic dynamical systems," intended for any level of readership at or above the undergraduate level, including non-mathematicians. We emphasize the beautiful relationship between fractal functions (such as Weierstrass's) and chaotic dynamical systems; these nowhere-differentiable functions are generating functions of chaotic dynamical systems. We can show that these functions are in a sense unique solutions of certain boundary problems. We can generalize to multiple dimensions. The last chapter of this book treats harmonic functions on fractal sets.

Library of Congress Cataloging-in-Publication Data

Yamaguchi, Masaya, 1925–
    [Furakutaru no sūri. English]
    Mathematics of fractals / Masaya Yamaguti, Masayoshi Hata, Jun Kigami ; translated by Kiki Hudson.
        p. cm. — (Translations of mathematical monographs ; v. 167)
    Includes bibliographical references (p. –) and index.
    ISBN 0-8218-0537-1 (acid-free paper)
    1. Fractals. 2. Chaotic behavior in systems. I. Hata, Masayoshi. II. Kigami, Jun. III. Title. IV. Series.
QA614.86.Y3513   1997
514′.742—dc21   97-11489
                                                                                                                     CIP

Contents

Preface

Studies of fractals in Japan began in the 1980's. In Chapters One and Two Masayoshi Hata discusses Mandelbrot's fractals (or definite fractals), using his own work on self-similarities. Masaya Yamaguchi explains, in Chapter Three, one method of computing fractal functions in place of Newtonian differentiation, together with wavelet analysis. This is a joint work with Masayoshi Hata. In the final chapter Jun Kigami explains a tentative development of fractal analysis.

Although the reader may find much of the material covered in Chapters One and Two in textbooks elsewhere, the way we explained them is original, reflecting our research philosophy. The reader will not easily find what we offer in Chapters Three and Four anywhere else, for the contents of these chapters, both materials and methods, are our own.

We are very happy to have an opportunity to ask a wider audience to judge our book. We are deeply indebted to the members of the editorial board of this lecture series and the editors at the Iwanami Shoten Publishing Company.

Masaya Yamaguti

Masayoshi Hata

Jun Kigami

January, 1993

Preface to the English Translation

It is a great honor for the authors that their book "Mathematics of Fractals" was selected for translation into English by the American Mathematical Society. We hope that this will help a new field of mathematics to be accepted by a broader audience.

This book aims at providing a handy explanation of the notions behind the self-similar sets called "fractals" and "chaotic dynamical systems," intended for any level of readership at or above the undergraduate level, including non-mathematicians. The novel contributions presented here include the consideration of fractal functions as unique solutions of some boundary-value problems. We emphasize the beautiful relationship between some fractal functions (such as Weierstrass's) and chaotic dynamical systems; these nowhere-differentiable functions are generating functions of some chaotic dynamical systems. We can show that these functions are in some sense unique solutions of some boundary value problems. We can generalize to multiple dimensions. The last chapter of this book treats harmonic functions on some fractal sets.

After the publication of the original version of this book in 1993, the authors have moved to different fields: M. Hata to number theory, J. Kigami to analysis on fractal sets, and M. Yamaguti to the discretization of ordinary differential equations.

We take the opportunity here to thank the translator, K. Hudson, who helped us to make some improvements from the original Japanese-language version.

Masaya Yamaguti
Masayoshi Hata
Jun Kigami

February 1997

CHAPTER 1

The Fundamentals of Fractals

In the long history of mathematics, we have seen a handful of peculiar sets and functions, well beyond ordinary imagination, that strongly impacted the mathematical environments of their times. The Cantor ternary set, the Koch curve, and Weierstrass's nowhere-differentiable continuous function come to mind as examples of such occasions. Although such objects had been constructed as sporadic counter–examples, until quite recently there had hardly been any unified discussion of them and no one imagined that they would be useful as models for natural phenomena. Mandelbrot[1], noticing that these sets shared some common properties, called them fractals and produced computer drawings of various fractals which drew much attention from every walk of research, even outside mathematics. In this chapter we first introduce the Hausdorff measure and the Hausdorff dimension as basic vital facts necessary for understanding fractals. We then look at some fractals, including the ones mentioned above, as well as some singular functions.

1.1. What is the dimension?

Sets such as a point, a line or the interior of a square are mathematical objects which we can intuitively understand with common sense, and we have no trouble in saying that their dimensions are zero, one or two respectively. But assigning a dimension to an arbitrary set (here we consider, for simplicity, only subsets of the real line, of the plane or more generally of n-dimensional Euclidean space $\mathbf{R}^n$) in a natural way–we call this correspondence a dimension function–is never a trivial endeavor. We can say the same about other intuitive concepts such as areas or volumes, which are not God-given and therefore require rigorous definition.

Let us clarify what we mean by the natural correspondence in the above paragraph. For an arbitrary set $X \subset \mathbf{R}^n$ we require that the dimension of X, written $\dim(X)$, satisfy the following properties:

(1) For the singleton set $\{p\}$, $\dim(\{p\}) = 0$, for the unit interval I^1, $\dim(I^1) = 1$, and in general, for the m-dimensional hypercube I^m, $\dim(I^m) = m$.

(2) (*Monotonicity*)   If $X \subset Y$,

$$\dim(X) \leq \dim(Y).$$

(3) (*Countable stability*)   If $\{X_j\}$ is a sequence of closed subsets of $\mathbf{R}^n$, then

$$\dim\left(\bigcup_{j=1}^{\infty} X_j\right) = \sup_{j \geq 1} \dim(X_j).$$

---

[1]Mandelbrot, B. B., *The Fractal Geometry of Nature*, W. H. Freeman, San Francisco, 1982.

(4) (*Invariance*) For an arbitrary map ψ belonging to some subfamily of the set of homeomorphisms of $\mathbf{R}^n$ to $\mathbf{R}^n$

$$\dim(\psi(X)) = \dim(X).$$

Of the above conditions (1) and (2) are fairly natural requirements. The condition (3) may seem slightly technical; however, replacing "countably many" by "finitely many" we have the following *finite stability* which is quite reasonable:

(3′) If $X_1, X_2, \ldots, X_m$ are closed sets in $\mathbf{R}^n$,

$$\dim \left(\bigcup_{j=1}^{m} X_j \right) = \max_{1 \leq j \leq m} \dim(X_j).$$

The condition (3) is simply an extension of (3′) which reduces to the case $m = 2$. The condition (4) is very important in that, for example, it is inevitable to ask that the dimension of a set X agree with the dimension of the set X' which is a parallel transform of X. In general, it is reasonable to require the invariance under congruent transformations, similarity transformations and so forth. Thus the subfamily of the condition (4) must contain at least congruent transformations and similarity transformations. In particular we see from the conditions (1), (2) and (4) that open sets in $\mathbf{R}^n$ are n-dimensional.

When we insist on the condition (4) for every homeomorphism we say that the dimension function under consideration is *topologically invariant*. Here we ask that sets which are homeomorphic to each other have the same dimension. It is not trivial to see that the conditions (1) and (4) do not contradict each other. Historically, toward the end of the nineteenth century, Peano constructed a continuous map from the unit interval $[0, 1]$ onto the square $[0, 1] \times [0, 1]$, astounding the mathematicians of the time.

Luckily, however, we know that the Peano map cannot be made homeomorphic: it is impossible to make the Peano map one-one. Urysohn and others actually constructed a topologically invariant dimension function which satisfies the above conditions and which takes on integer values. We write $\dim_T$ for this function and call it the *topological dimension*. Its construction is based on the idea of generalizing the notion that the dimension of a ball is three and the dimension of the sphere which is the surface of the ball is two, to define inductively the dimension of a set X from the dimension of its boundary ∂X. To be precise: in the theory of general topology, we have three kinds of dimensions, two inductive dimensions and a covering dimension, all of which agree in a countable metric space. Since we are dealing with subsets of Euclidean spaces we may call any of these dimension functions the topological dimension. We also mention that topologically invariant dimension functions which take values on integers are not unique. For more details we refer the reader to the classic of dimension theory by Hurewicz and Wallman[2].

On the other hand, mathematicians had for a long time made unsophisticated but important attempts to generalize the intuitive ideas such as length, area and volume on more general sets. For example, Borel initiated his research in this direction at the end of the nineteenth century, and this stimulated Lebesgue's work on measure theory as well as integral theory at the beginning of the twentieth century. Carathéodory generalized Lebesgue's work to s-dimensional measure in $\mathbf{R}^n$, and then Haufsdorff, noticing that s made sense even if it was not an integer,

---

[2]Hurewicz, W. and Wallman, H., *Dimension Theory*, Princeton Univ. Press, 1948.

came to define the Hausdorff dimension. The Hausdorff dimension satisfies all the dimension conditions with the set of bi-Lipschitz maps for the subfamily in condition (4). We denote by $\dim_H$ this real-valued dimension function, and discuss its definition and properties in the next section.

In this way we have come to possess two dimension functions, $\dim_T$ and $\dim_H$, which provide us with two measuring devices for the dimension of a set. Sometimes the two devices give the same measurement and other times their measurements differ; however, for any set X we have the inequality

$$(1.1) \qquad \dim_T(X) \le \dim_H(X);$$

therefore if they are different the Hausdorff dimension is always larger. Mandelbrot, recognizing this fact, gave the following

DEFINITION 1.1. We say that a set X in $\mathbf{R}^n$ is *fractal* if

$$\dim_T(X) < \dim_H(X).$$

The properties (1) and (4) for dimension-functions imply that familiar sets such as line segments and polygons in Euclidian geometry are not fractal; neither are open sets in $\mathbf{R}^n$.

REMARK. There are sets which are not fractal according to our definition but appear to be fractal according to our senses. Mandelbrot regards these sets as being on the border between the ordinary sets and the fractals. The graph G of the Takagi function, a well-known example of a nowhere differentiable continuous function, is such a set. In fact one can show that

$$\dim_T(G) = \dim_H(G) = 1$$

(Exercise 1.2); however, the fact that the function is not differentiable makes its graph G somewhat jagged, and so we should distinguish it from ordinary lines.

From Definition 1.1 one naturally defines *the fractal degree* of X for a set X, which shows how fractal X is, by

$$\delta(X) = \dim_H(X) - \dim_T(X).$$

As $\dim_H(X) \le n$, one has $0 \le \delta(X) \le n$. In particular, since $\dim_T(X)$ takes values on integers only, we have the following

THEOREM 1.1. *A set X is fractal if $\dim_H(X)$ takes on a non-integer value.*

We also get from (1.1) the following equality:

$$\dim_T(X) = \inf_{X \sim Y} \dim_H(Y),$$

where the inf on the right hand side runs through any Y homeomorphic to X.

We conclude this section by introducing one other real-valued dimension. For simplicity we let "ϵ-ball" mean a closed ball in $\mathbf{R}^n$ of radius ϵ. Let X be a bounded set in $\mathbf{R}^n$. For an arbitrary positive number ϵ, let $N(\epsilon)$ be the minimum number of ϵ-balls needed to cover X. Since X is bounded it has finite covers, and we look at the number of balls in each of them without worrying about how they cover X. In this setting we say that the following limit, if it exists,

$$\lim_{\epsilon \to 0_+} \frac{\log N(\epsilon)}{\log(1/\epsilon)}$$

is the *box-counting dimension* of X and denote it by $\dim_B$. Roughly speaking the magnitude of $N(\epsilon)$ is about ϵ^{-s}, where s denotes the above limit. For example, in the case of $I^1 = [0, 1]$ we have $N(\epsilon) \sim \epsilon^{-1}$, and we have $N(\epsilon) \sim \epsilon^{-2}$ for the square $I^2 = [0, 1] \times [0, 1]$; hence the box-counting dimensions of standard sets agree with their (usual) dimensions. In fact the function $\dim_B$ satisfies the dimension conditions when we replace the property (3) by the finite stability (3′) and require the invariance (4) with respect to the bi-Lipschitz maps.

The following example shows that the box-counting dimension does not satisfy the countable stability: in $\mathbf{R}^1$ set

$$X_0 = \left\{ 0, 1, \frac{1}{2}, \frac{1}{3}, \ldots, \frac{1}{n}, \ldots \right\}.$$

If $\dim_B$ is countably stable we must have $\dim_B(X_0) = 0$ as X_0 is a countable set, but we can easily show that $\dim_B(X_0) = 1/2$ (Exercise 1.3).

The box-counting dimension relates to other types of dimensions by the inequalities

(1.2) $$\dim_T(X) \leq \dim_H(X) \leq \dim_B(X);$$

we will show the second inequality in Section 1.2. Thus it is plausible to include a set X with

$$\dim_T(X) = \dim_H(X) < \dim_B(X)$$

in the fractal family in the sense that it possesses distinct dimensions; the set X_0 defined above provides such an example. However, whether X_0 behaves as a singular set depends on various circumstances, and the definition of fractals should be made accordingly.

1.2. Hausdorff measure and Hausdorff dimension

In this section we introduce the Hausdorff measure and the Hausdorff dimension, and discuss their basic properties. We denote by $\|.\|$ the usual norm on n-dimensional Euclidean space $\mathbf{R}^n$. By the *diameter* of a set U we mean

$$|U| = \sup_{x,y \in U} \|x - y\|.$$

A δ-*covering* of a given set X is a countable sequence of sets $\{U_i\}$ which satisfy

$$X \subset \bigcup_{i=1}^{\infty} U_i \quad \text{and} \quad 0 < |U_i| \leq \delta \qquad (i \geq 1).$$

This definition of course includes finite coverings, the most extreme case being that X is its own $|X|$-covering. The definition depends only on the diameter of each set in the δ-covering $\{U_i\}$, and has nothing to do with the way X is covered. For instance, if we replace each U_i with its closure $\overline{U}_i$, we still have a δ-covering because $|U_i| = |\overline{U}_i|$. Similarly we may consider only compact sets or only convex sets for our coverings. Now set

(1.3) $$\mathcal{H}_{\delta}^s(X) = \inf_{\{U_i\}} \sum_{i=1}^{\infty} |U_i|^s,$$

where s is a non-negative real number and the inf on the right hand side runs through the possible δ-coverings of X. For a fixed set X, $\mathcal{H}_{\delta}^s(X)$ is a function of two variables, s and δ. Notice, however, that if the series on the right hand side of

(1.3) diverges we take the value of the infimum to be ∞, *i.e.* the function $\mathcal{H}_\delta^s$ takes on the values in $[0, \infty]$. With a fixed s, in particular, $\mathcal{H}_\delta^s$ is a function of δ which increases monotonically as δ decreases, the reason being that a δ-covering $\{U_i\}$ is also a δ'-covering for any $\delta' \geq \delta$. Hence, when we include ∞ in its values we always have the limit

$$\mathcal{H}^s(X) = \lim_{\delta \to 0+} \mathcal{H}_\delta^s(X),$$

which satisfies the following conditions for an outer measure called the *s-dimensional Hausdorff–outer measure*:

(a) $\mathcal{H}^s(\emptyset) = 0$, where $\emptyset$ is the empty set.
(b) $\mathcal{H}^s(X) \leq \mathcal{H}^s(Y)$, if $X \subset Y$.
(c) For any sequence $\{X_j\}$ of subsets in $\mathbf{R}^n$,

$$\mathcal{H}^s\left(\bigcup_{j=1}^\infty X_j\right) \leq \sum_{j=1}^\infty \mathcal{H}^s(X_j).$$

Unlike the dimension functions introduced in Section 1.1, an outer measure is a set function which takes on values in $[0, \infty]$. We mention, without going into detail, that a Borel set is measurable with respect to the outer measure $\mathcal{H}^s$. Here we say that a set is Borel if it is obtained through repeated operations of taking the complement of a closed set and forming a countable union of sets in $\mathbf{R}^n$. In particular, we say that a union of countably many closed sets is an F_σ-set and an intersection of countably many open sets is a G_δ-set. In other words, if the X_j's in (c) are pairwise disjoint and each X_j is Borel, then we have:

(c′) (complete additivity) $\mathcal{H}^s\left(\bigcup_{j=1}^\infty X_j\right) = \sum_{j=1}^\infty \mathcal{H}^s(X_j).$

It follows from the definition that the values of $\mathcal{H}^s$ are invariant under congruent transformations of $\mathbf{R}^n$. The concepts of length, area and volume extend to $\mathcal{H}^1$, $\mathcal{H}^2$ and $\mathcal{H}^3$ respectively. More precisely, $\mathcal{H}^n$ is a constant multiple of the n-dimensional Lebesgue measure. In particular, when $n = 1$ they agree completely.

Whereas $\mathcal{H}^s(X)$ is an outer measure as a function of X, it can be regarded as a monotone decreasing function of s with X fixed, because for any $s < t$ and any δ-covering $\{U_i\}$ of X we have

$$\sum_{i=1}^\infty |U_i|^t = \sum_{i=1}^\infty |U_i|^s |U_i|^{t-s} \leq \delta^{t-s} \sum_{i=1}^\infty |U_i|^s,$$

and thus $\mathcal{H}_\delta^t(X) \leq \delta^{t-s} \mathcal{H}_\delta^s(X)$. In particular, if $\mathcal{H}^s(X) < \infty$ we get $\mathcal{H}^t(X) = 0$, and if $\mathcal{H}^t(X) > 0$ we must have $\mathcal{H}^s(X) = \infty$. In other words, the graph of $\mathcal{H}^s(X)$ is a step function with at most one point of discontinuity (of the first kind). See Figure 1.1.

We call the value of this discontinuity point d, if it exists, the *Hausdorff dimension* of X and write $\dim_H(X)$. From Figure 1.1 this value evidently satisfies the equalities

$$\dim_H(X) = \sup_s\{\mathcal{H}^s(X) = \infty\} = \inf_s\{H^s(X) = 0\}.$$

The value of $\mathcal{H}^s(X)$ at the discontinuity point $s = d$ may be 0, a positive number, or ∞; in case $0 < \mathcal{H}^d(X) < \infty$ we say that X is a *d-set*. The situation $\mathcal{H}^s(X) = \infty$ for $s < d$ means that the ruler $\mathcal{H}^s$ is so fine that X looks too big to

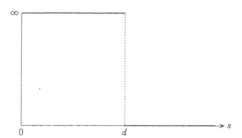

FIGURE 1.1. The graph of $\mathcal{H}^s$ as function of s

be measured. On the other hand, $\mathcal{H}^s(X) = 0$ for $s > d$ means that the ruler $\mathcal{H}^s$ is too coarse to measure X, which looks too small to notice. In this sense, when X is a d-set, $\mathcal{H}^d$ is calibrated just right to measure X.

In conjunction with the dimension property (4) we show the following:

LEMMA 1.1. *Let A be a subset of $\mathbf{R}^n$ and $\psi : A \to \mathbf{R}^n$ a Lipschitz map, that is, for some constant c, $\|\psi(x) - \psi(y)\| \le c\|x - y\|$ for all $x, y \in A$. Then for an arbitrary $s \ge 0$, we have $\mathcal{H}^s(\psi(A)) \le c^s \mathcal{H}^s(A)$. In particular, we have*

$$\dim_H(\psi(A)) \le \dim_H(A).$$

PROOF. Pick any δ-covering for A. Evidently the sequence $\{\psi(A \cap U_i)\}$ covers $\psi(A)$ in such a way that

$$|\psi(A \cap U_i)| = \sup_{x,y \in A \cap U_i} \|\psi(x) - \psi(y)\| \le c \sup_{x,y \in U_i} \|x - y\| = c|U_i|;$$

so it is a $c\delta$-covering of $\psi(A)$. Hence,

$$\mathcal{H}^s_{c\delta}(\psi(A)) \le \sum_{i=1}^{\infty} |\psi(A \cap U_i)|^s \le c^s \sum_{i=1}^{\infty} |U_i|^s.$$

From these inequalities we get $\mathcal{H}^s_{c\delta}(\psi(A)) \le c^s \mathcal{H}^s_{c\delta}(A)$ as we chose the $\{U_i\}$ arbitrarily. Taking the limit as $\delta \to 0$ on each side, we get $\dim_H(\psi(A)) \le \dim_H(A)$. $\square$

We can generalize Lemma 1.1 to the following:

LEMMA 1.2. *Let A be a subset of $\mathbf{R}^n$ and $\psi : A \to \mathbf{R}^n$ a Lipschits map whose degree is $\alpha \in (0, 1]$, that is, for some constant c, $\|\psi(x) - \psi(y)\| \le c\|x - y\|^\alpha$ for any $x, y \in A$. Then $\dim_H(\psi(A)) \le \alpha^{-1} \dim_H(A)$.*

We omit the proof, which is similar to that of Lemma 1.1.

In the case when the map $\psi : \mathbf{R}^n \to \mathbf{R}^n$ is a bi-Lipschitz homeomorphism, that is, for some constants c_1 and c_2 the inequalities

$$c_1\|x - y\| \le \|\psi(x) - \psi(y)\| \le c_2\|x - y\|$$

hold for every pair $x, y \in \mathbf{R}^n$, it follows immediately from Lemma 1.1 that

$$\dim_H(\psi(A)) = \dim_H(A).$$

In other words, the invariance property (4) holds for bi-Lipschitz maps. For an application of Lemma 1.1, let proj be the orthogonal projection of $\mathbf{R}^n$ onto an r-dimensional subspace. Then, as proj is a map which does not increase the distance, we get

(1.4) $\dim_H(\mathrm{proj}(A)) \leq \dim_H(A).$

We conclude this section by showing the second inequality of (1.2), which relates the Hausdorff dimension and the box–counting dimension. Consider $N(\epsilon)$ many ϵ-balls covering a set $X \in \mathbf{R}^n$. Evidently they form an ϵ-covering of X, and so we have

$$\mathcal{H}_\epsilon^s(X) \leq N(\epsilon)\epsilon^s.$$

We note now that $\mathcal{H}^s(X) = \infty$ for arbitrary $s < \dim_H(X)$, so that $\mathcal{H}_\epsilon^s(X) > 1$ for any sufficiently small $\epsilon > 0$. Hence we get, taking the logarithm on both sides of the inequality,

$$0 < \log \mathcal{H}_\epsilon^s(X) \leq \log N(\epsilon) + s \log \epsilon;$$

that is,

$$s \leq \frac{\log N(\epsilon)}{\log(1/\epsilon)}.$$

Since we can make ϵ as small as we desire, we get $s \leq \dim_B(X)$ if the right–hand side of the above inequality has a limit. On the other hand, as we can take s arbitrarily close to $\dim_B(X)$, we may conclude that $\dim_H(X) \leq \dim_B(X)$.

1.3. Examples of fractals and their Hausdorff dimensions

In this section we give basic examples of fractal sets whose Hausdorff measures and Hausdorff dimensions are calculable directly from the definitions. In general it is not easy to determine even the Hausdorff dimension of a set X in $\mathbf{R}^n$. The difficulty arises from the fact that whereas we can easily estimate $\dim_H(X)$ from above by choosing some suitable δ-covering, we have to consider the entire family of coverings of X to estimate it from below.

The situation in one-dimensional cases, however, is relatively simple. We noted in the previous section that the definition of a δ-covering did not lose its generality if we use a sequence $\{U_i\}$ of closed convex sets as a covering, so that it is enough to consider sequences of closed intervals of length less than or equal to δ to cover a subset $X \subset \mathbf{R}^1$.

We start with the unit interval $[0, 1]$, which we denote by C_0. We remove the open interval of length $1/3$ from the center of C_0, and we denote the remaining set by C_1, $C_1 = [0, 1/3] \cup [2/3, 1]$. We continue the process of removing from the center of each newly created subinterval the open interval whose length is one-third of the subinterval to define inductively the kth set C_k; C_k is a union of 2^k subintervals of length 3^{-k}, and $\{C_k\}$ is a monotone decreasing sequence of compact sets. See Figure 1.2.

The limit of this sequence

$$C = \bigcap_{k=1}^\infty C_k$$

is a compact set called the *Cantor set* or Cantor's ternary set. This set looks porous even to the naked eye. More precisely, the connected component at each point of C is the singleton set of the point itself; we say that C is *totally disconnected*.

The set C is a complete set; that is, C is a closed set containing no isolated points. Now there exists a ball centered at each point of C with arbitrarily small radius such that its circumference does not intersect C. Hence it follows by the inductive definition of the topological degree in §1.1 that $\dim_T(C) = 0$.

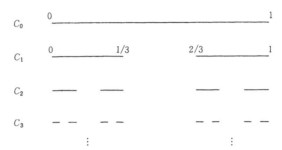

FIGURE 1.2. The construction of the Cantor set

Furthermore, the following theorem implies that the Cantor set is fractal. The Cantor set is one of the rare sets whose Hausdorff measures are accurately calculable. We can also determine the Hausdorff dimension of C. In the next chapter we discuss *self-similar sets* in detail and, in particular, show that we can determine the Hausdorff dimensions of similar sets which satisfy certain properties. The Cantor set is a typical example of such a set.

THEOREM 1.2. *The Hausdorff dimension s of the Cantor set C is $\log 2/\log 3$ $(= 0.63092\cdots)$, and the s-dimensional Hausdorff outer measure of C is one; that is, $\mathcal{H}^s(C) = 1$.*

PROOF. It is clear that C_k is a δ-covering of C as $C \subset C_k$, and that C_k is the union of 2^k many subintervals of length δ, where $\delta = 3^{-k}$. Hence

$$\mathcal{H}^s_\delta(C) \le 2^k \cdot 3^{-sk} = 1,$$

and as we can make δ as small as necessary by taking k sufficiently large, we have that $\mathcal{H}^s(C) \le 1$.

We next show the reverse inequality. We can show by contradiction a stronger statement: for any covering $\{I_i\}$ of C by closed intervals, the inequality

$$\sum_{i=1}^\infty |I_i|^s \ge 1$$

holds. This clearly implies that $\mathcal{H}^s(C) \ge 1$.

Let us assume now that there is a covering $\{I_i\}$ of C with $\sum |I_i|^s < 1$. We may assume that each I_i is an open interval, since we can expand each I_i slightly while keeping the inequality; then the compactness of C implies that C is actually covered by finitely many I_i's, and so we may select without loss of generality a finite open covering $\{I_i\}$. So suppose that $C \subset \bigcup_{i=1}^m I_i$. Then for a large enough k, we have

$$C_k \subset \bigcup_{i=1}^m I_i,$$

and we choose k to be the smallest number with the property that each subinterval of length 3^{-k} comprising C_k is contained in some I_i. We consider the case where each I_i contains at least two subintervals. The set $I_i \backslash C_k$ – the subset of I_i obtained by removing the points belonging to C_k – is a union of finitely many open intervals.

Let E be one of these intervals with the maximum length. We may assume that the end points of E are in the interior of I_i by shrinking I_i slightly as needed.

If for simplicity we set $I_i = E \cup E_0 \cup E_1$, where E_0 and E_1 are the resulting half-intervals of removing E from I_i, then the construction of C_k implies that $|E_0|, |E_1| \leq |E|$. Thus we get

$$|I_i|^s = (|E| + |E_0| + |E_1|)^s \geq \left(\frac{3}{2}(|E_0| + |E_1|) \right)^s$$

$$\geq 2 \left(\frac{|E_0| + |E_1|}{2} \right)^s \geq |E_o|^s + |E_1|^s.$$

Here we used the convexity of the function $f(x) = x^s$. These inequalities show that if we replace the covering $\{I_i\}$ by $E_0 \cup E_1$, the sum of their respective s-th powers is still less than or equal to one. Hence we may assume that each I_i has exactly one subinterval of length 3^{-k}. In this case we have

$$\sum_{i=1}^m |I_i|^s \geq 2^k \cdot 3^{-sk} = 1,$$

which is a contradiction. This completes the proof of the theorem. $\qquad \square$

The Cantor set is called ternary because it has the expression

$$C = \left\{ x \in [0,1]; \ x = \sum_{n=1}^{\infty} \frac{a_n}{3^n}, \ a_n \in \{0, 2\} \right\};$$

in other words, C agrees with the set of real numbers in the interval $[0,1]$ whose ternary expressions never contain the numeral 1. We point out, however, that for example $1/3$ has an expression $1/3 = 0.1000 \cdots (3)$ as well as $1/3 = 0.0222 \cdots (3)$; so this type of number does belong to C.

The Cantor set has the following curious property: *For an arbitrary real number* $x \in [-1, 1]$, *there exist numbers* $y, z \in C$ *such that* $x = y - z$. We indicate this fact by $[-1, 1] \subset C - C$. Notice on the other hand that the Lebesgue measure of the Cantor set is zero, and compare the above fact with Steinhaus's famous theorem: *If the Lebesgue measure of A is positive then $A - A$ contains some neighborhood of the origin.*

The set L of Liouville numbers is even thinner but satisfies the above condition. One had known that the Hausdorff dimension of L is zero before Erdös[3] proved the fact that $L - L = \mathbf{R}$. We remind the reader that while the set L is of the second category in the sense of Baire, the Cantor set is of the first category.

Using the Cantor set, we construct the set $X = (C \times I) \cup (I \times C)$ (Figure 1.3), where I is the unit interval $I = [0, 1]$, in the plane. Despite the fact that the two-dimensional Lebesgue measure of X is 0, X contains the sides of a rectangle of lengths a and b for arbitrary real numbers $a, b \in (0, 1]$.

We next introduce an example of a totally disconnected set in the plane $\mathbf{R}^2$, whose Hausdorff dimension is easy to compute.

Set $S_0 = [0, 1] \times [0, 1]$ as a unit square in $\mathbf{R}^2$ and repeat the operation consecutively as indicated in Figure 1.4 to construct the set S_k. The set S_k consists

---

[3]Erdös, P., Representations of real numbers as sums and products of Liouville numbers, Michigan Math. J. 9(1962), 59–60.

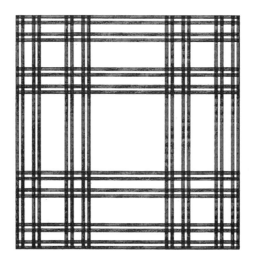

FIGURE 1.3. The set $(C \times I) \cup (C \times I)$

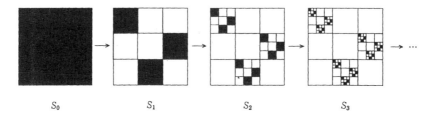

S_0 S_1 S_2 S_3

FIGURE 1.4. Construction of the set S

of 3^k squares each having the width 3^{-k}, and the set sequence $\{S_k\}$ is monotone decreasing. We look at a compact set S defined by

$$S = \bigcap_{k=1}^{\infty} S_k.$$

Just like the Cantor set, S is totally disconnected and its topological dimension is zero. Notice that S_k itself is a $\delta = 3^{-k}$-covering of S, so that

$$\mathcal{H}_\delta^1(S) \le 3^k \cdot 3^{-k} = 1,$$

where we can make δ as small as we desire. Hence $\mathcal{H}^1(S) \le 1$, and $\dim_H(S) \le 1$.

We next look at the orthogonal projection, $\mathrm{proj}(S)$, of S to the x-axis. Here we get $\mathrm{proj}(S) = [0, 1]$ for the following reason. For any real number $x \in [0, 1]$, let l_x be a line through the point $(x, 0)$ parallel to the y-axis. Then $S_k \cap l_x \ne \emptyset$ for each k, and we have a monotone decreasing sequence of compact sets $\{S_k \cap l_x\}$; therefore, we get the equality

$$S \cap l_x = \bigcap_{k=1}^{\infty} (S_k \bigcap l_x) \ne \emptyset,$$

which implies that $x \in \mathrm{proj}(S)$. It follows from the property (1.4) of the Hausdorff dimension that

$$1 = \dim_H([0,1]) = \dim_H(\mathrm{proj}(S)) \leq \dim_H(S),$$

and we end up with $\dim_H(S) = 1$.

1.4. Nowhere-differentiable functions

In 1872, Weierstrass constructed the first example of a continuous function which is nowhere differentiable:

$$(1.5) \qquad\qquad W(x) = \sum_{n=0}^{\infty} a^n \cos(b^n \pi x),$$

where $0 < a < 1$ and b is an odd integer with $ab > 1 + 3\pi/2$. In other words, one cannot draw the tangent line at any point of the graph of $W(x)$. The story goes that this absolutely convergent function $W(X)$ given in a simple Fourier series astounded the mathematical circle of the time. The function must have puzzled mathematicians because while each of its terms is smooth it becomes non-differentiable when all are put together.

After Weierstrass the first essential progress came with Hardy[4], who proved that $W(x)$ had no bounded derivatives if $ab \geq 1$. If $ab < 1$ we see by termwise differentiation that $W(x)$ is continuously differentiable, and so $ab = 1$ is the dividing line for differentiability and non-differentiability. For example, Hata[5] showed that if b is a real number, not necessarily an odd integer, which satisfies $ab \geq 5.603 \cdots$ then $W(x)$ has no derivatives including ∞.

Figure 1.5 shows the graph of $W(x)$ for the typical case $b = 1/a = 2$ (it is actually the graph of $1/2(1 - W(x))$ to be compared with the graph of the Takagi function). Unfortunately Hardy's proof of the non-differentiability of $W(x)$ is almost too technical and too complicated here. So we replace the trigonometric function $\cos(\pi x)$ by the function $\psi(x) = \mathrm{dist}(x, \mathbf{Z})$, the distance between x and the integer closest to x, whose graph looks like saw-teeth as shown in Figure 1.6, and consider the function

$$T(x) = \sum_{n=0}^{\infty} \frac{1}{2^n} \psi(2^n x).$$

Then T is also a continuous function everywhere without bounded derivative; however, the proof of this fact is much simpler. This is the *Takagi function* (Figure 1.7) discovered in 1903 by Takagi[6].

The graghs of the Takagi and Weierstrass functions are quite similar. We can see that the graph of Takagi's function has self-similarity; that is, it consists of parts which are similar contractions of the whole. It is interesting to note that

---

[4]Hardy, G. H., Weierstrass's non-differentiable function, Trans. Amer. Math. Soc., 17(1916), 301–325

[5]Hata, M., Singularities of Weierstrass type functions, Journal d'Analyse Mathématique, vol. 51(1988), 62–90.

[6]Takagi, T., A simple example of the continuous function without derivative, The Collected Papers of Teiji Takagi, Iwanami Shoten Publ., Tokyo, 1973, 5-6.

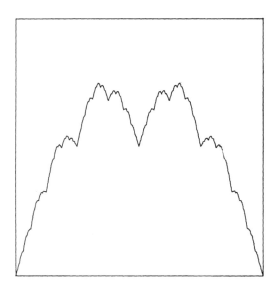

FIGURE 1.5. The Weierstrass function for $b = 1/a = 2$

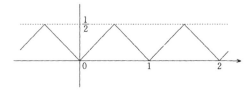

FIGURE 1.6. The graph of $\psi(x) = \mathrm{dist}(x, \mathbf{Z})$

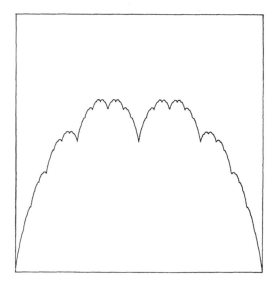

FIGURE 1.7. The graph of the Takagi function

Knopp[7] (1918), Hobson[8](1926), van der Waerden[9] (1930), Hildebrandt[10] (1933), de Rham[11] (1957), and other mathematicians have repeatedly rediscovered functions essentially identical to the Takagi function.

The function $T(x)$ goes further than just being a simplification of the Weierstrass function (1.5). We shall see in Chapter Three that $T(x)$ is relevant to the Schauder expansion as well as Lebesgue's singular functions (strictly increasing functions whose derivatives are almost everywhere zero), thus offering us many interesting problems.

Intuitively we imagine that the graph of a non-differentiable function such as Weierstrass's or Takagi's function is bumpy and quite oscillatory in any subinterval however small. Thus we come up with the interesting problem of observing the Hausdorff dimension of the graph of a continuous function $f(x)$,

$$G_f = \{ (x, f(x)) \, ; \, x \in [a, b] \} \, ,$$

as a subset of the plane. The graph of a continuous function is a continuous curve without self-intersections, and so its topological degree is one. Thus if we show that $\dim_H(G_f) > 1$ then we will know that G_f is a fractal, and in the case where f is nowhere differentiable we will have justified our intuitive image of its graph as discussed above.

We begin by evaluating from above the Hausdorff dimension of the graph of a given continuous function.

THEOREM 1.3. *Let a continuous function $f(x)$ defined on the interval $[a, b]$ be Lipschitz of degree $\alpha \in (0, 1]$ (i.e., there exists a constant c such that*

$$|f(x) - f(y)| \leq c|x - y|^\alpha$$

for any $x, y \in [a, b]$). Then $\mathcal{H}^{2-\alpha}(G_f) < \infty$, and $\dim_H(G_f) \leq 2 - \alpha$ in particular.

PROOF. Divide the interval $[a, b]$ into k equal parts and label them from the left, $I_1, I_2, \ldots, I_k$. Note that $|I_i| = (b - a)/k$. Now by moving $y \in I_i$ around while keeping the left endpoint x of I_i fixed, we get

$$|f(x) - f(y)| \leq c|x - y|^\alpha \leq c|I_i|^\alpha = c(b - a)^\alpha k^{-\alpha}.$$

Hence, if starting at the point $(x, f(x))$ we stack $\left[c(b - a)^{\alpha-1}k^{1-\alpha}\right]+1$ many squares of width $(b-a)/k$ up and stack the same number of them down, the point $(y, f(y))$ will belong to one of them. The family of these squares taken over the entire interval $[a, b]$ is a δ-covering of G_f, where $\delta = \sqrt{2}(b - a)/k$ is the diameter of any square

[7]Knopp, K., Ein einfaches Verfahren zur Bildung stetiger nirgends differenzierbarer Funktionen, Math. Z., 2(1918), 1–26.

[8]Hobson, E. W., *The Theory of Functions of Real Variables and the Theory of Fourier Series*, vol. II, Harren, Washington, 1950.

[9]van der Waerden, B. L., Ein einfaches Beispiel einer nichtdifferenzierbaren stetigen Function, Math. Z., 32(1930), 474–475.

[10]Hildebrandt, T. H., A simple continuous function with a finite derivative at no point, Amer. Math. Monthly, 40(1933), 547–548.

[11]de Rham, G., Sur exemple de fonction continue sans dérivée, Enseign. Math., 3(1957), 71–72.

under consideration; therefore, we get

$$\mathcal{H}_\delta^{2-\alpha}(G_f) \le 2k \left(\frac{\sqrt{2}(b-a)}{k} \right)^{2-\alpha} \left\{ 1 + c(b-a)^{\alpha-1}k^{1-\alpha} \right\}$$

$$\le 2 \left(\sqrt{2}(b-a) \right)^{2-\alpha} \left\{ 1 + c(b-a)^{\alpha-1} \right\},$$

where the right-hand side is a constant independent of k. We may also make δ arbitrarily small, and so it follows that $\mathcal{H}^{2-\alpha}(G_f) < \infty$.   □

When the numbers a and b satisfy $ab > 1$, Weierstrass's function (1.5) is Lipschitz of degree $\log(1/a)/\log b$ (Exercise 1.10), and hence Theorem 1.3 implies

$$\dim_H(G_W) \le 2 - \frac{\log(1/a)}{\log b}.$$

Here one speculates in general that the equality might hold; however, to obtain an accurate value of the Hausdorff dimension of $W(x)$ would be quite a difficult undertaking. Recently Mauldin and Williams[12] showed that

$$\dim_H(G_W) = 2 - \frac{\log(1/a)}{\log b} + O\left(\frac{1}{\log b} \right),$$

which might lend a slight tinge of plausibility to the above conjecture.

Furthermore, Przytycki and Urbański[13] showed that for arbitrary $ab > 1$ we get $\dim_H(G_W) > 1$, so that the graph of Weierstrass's function turns out to be a fractal. We note that if $ab = 1$ then $\dim_H(G_W) = 1$, as is the case for Takagi's function.

Exercises

1.1. Show that a monotone continuous function is not fractal.

1.2. Show that the Hausdorff dimension of the graph of Takagi's function $T(x)$ is one. [Hint: Show that $|T(x) - T(x+h)| = O(|h| \log(1/|h|))$.]

1.3. Show that the box-counting dimension of the set

$$X_0 = \left\{ 0, 1, \frac{1}{2}, \frac{1}{3}, \dots, \frac{1}{n}, \dots \right\}$$

is $1/2$. [Hint: For the interval I consider a natural number k which satisfies $1/(k^2 - k) > |I| \ge 1/(k^2 + k)$.]

1.4. Show that $[-1,1] \subset C - C$, where C is the Cantor set. [Hint: Use the set product $C_k \times C_k$.]

1.5. Show that there exists a continuous function whose graph is of Hausdorff dimension two. [Hint: Use the result of Mauldin and Williams, and the countable stability property of the dimension.]

1.6. Show that the Hausdorff dimension of the set product $C \times C$ of the Cantor set C is at least one.

---

[12]Mauldin, R. D., and Williams, S. C., On the Hausdorff dimension of some graphs, Trans. Amer. Math. Soc., 298(1986), 793–803.

[13]Przytycki, F., and Urbański, M., On the Hausdorff dimension of some fractal sets, Studia Math., 93(1989), 155–186.

1.7. Let V be a closed convex subset of the plane $\mathbf{R}^2$, and define a map $\psi : \mathbf{R}^2 \to V$ by

$$\psi(p) = \begin{cases} \text{the point on } \partial V \text{ closest to } p, \text{ if } p \notin V, \\ \text{the point } p, \text{ if } p \in V. \end{cases}$$

Show that $\dim_H(\psi(A)) \leq \dim_H(A)$ for any $A \subset \mathbf{R}^2$.

1.8. Show that the topological dimension of the graph of a continuous function is one.

1.9. Show that if $A \subset \mathbf{R}^n$ is bounded, then A and the closure of A have the same box-counting dimension.

1.10. Show that the Weierstrass function (1.5) is a Lipschitz map whose degree is $\log(1/a)/\log b$.

CHAPTER 2

Self-Similar Sets

A self-similar set, simply put, is a set which consists of miniatures of itsef. Self-similar sets among the fractals also form an important class, since many of them have accurately computable Hausdorff dimensions. We begin this chapter with discussions of the basic properties of self-similar sets together with various results on their sizes (Hausdorff dimensions) and shapes (connectivity and so forth). We then narrow our attention to a subclass of the self-similar sets consisting of self-affine sets, which are defined by affine contractions, and try to classify them by introducing basic relations among the fixed points of contractions. Self-affine sets come in so many different forms that their diversity never fails to amaze us. We will close the chapter with a brief dicussion of the relationship between chaos and fractals.

2.1. Existence and uniqueness

We can of course find the notion of self-similar sets in Mandelbrot's book mentioned in the previous chapter, but we cannot over-emphasize the interesting fact that a large number of counter-examples and singular sets that have appeared in mathematics possess some self-similarity trait in one sense or another. The Cantor set and the graph of the Takagi set are two examples of such sets. It goes without saying that the contribution of Mandelbrot here is great, as he is the one who noticed the self-similarity behind these individual examples and brought it out to a broader research domain.

It was Hutchinson[1] who initiated the general mathematical studies of self-similar sets, and a few years later Hata[2] developed independently a similar mathematical formulation. It turns out, however, that ten years prior to the work of Hutchinson, Williams[3] had already dealt with essentially the same sets, although he did not call them self-similar. His ideas, moreover, are quite intriguing in that they are based on the inverse maps of chaotic dynamical systems. We first collect the necessary definitions.

We say that the map $\psi : \mathbf{R}^n \to \mathbf{R}^n$ is a *contraction* if there exists some constant number $c \in (0, 1)$ so that the inequality

$$\|\psi(x) - \psi(y)\| \leq c\|x - y\|$$

holds for any $x, y \in \mathbf{R}^n$. In other words, a map is a contraction if it satisfies the definition of being Lipschitz (of degree one) with a constant c less than one. We

---

[1]Hutchinson, J. E., Fractals and self-similarity, Indiana Univ. Math. J., 30 (1981), 713–747.

[2]Hata, M., On the structure of self-similar sets, Japan J. Appl. Math., **2** (1985), 381–414.

[3]Williams, R. F., Composition of contractions, Bol. Soc. Brasil. Mat., **2** (1971), no. 2, 55–59.

define the *Lipschitz constant* of ψ to be the smallest among all such c's, and denote it by $L(\psi)$.

According to the principle of contraction in a complete metric space, ψ has a unique *fixed point* x; that is, there exists a unique point x which satisfies the equation

$$(2.1) \qquad\qquad\qquad x = \psi(x).$$

We simply write $F(\psi)$ for this fixed point. Now we extend the equation (2.1) to several contractions and make the following definition.

DEFINITION 2.1. Let $m \geq 2$ be a natural number and let $\{\psi_1, \psi_2, \ldots, \psi_m\}$ be a set of m contractions defined on $\mathbf{R}^n$. We say that a nonempty compact set V in $\mathbf{R}^n$ is *self-similar* if it satisfies

$$(2.2) \qquad\qquad\qquad V = \bigcup_{i=1}^{m} \psi_i(V).$$

We remind the reader that although we use the term *similar* in the above definition, we do not demand that each contraction be similar in the geometrical sense. These are just general contractions. However, Hutchinson's definition of self-similarity is quite a bit stricter, and he insists not only that the contractions be similar but also that they satisfy a certain separation condition, namely that the miniatures be mutually separated.

In fact we may regard (2.2), just like the equation (2.1), as the fixed point of a certain contraction. Our reasoning goes as follows. Let $K(\mathbf{R}^n)$ be the set of all non-empty compact sets in $\mathbf{R}^n$. For an element $A \in K(\mathbf{R}^n)$ we set

$$N_\epsilon(A) = \left\{ x \in \mathbf{R}^n;\ \mathrm{dist}(x, A) \equiv \min_{y \in A} |x - y| \leq \epsilon \right\},$$

and say that $N_\epsilon(A)$ is the ϵ-neighborhood of A. The set $K(\mathbf{R}^n)$ becomes an abstract metric space when we give it the following *Hausdorff metric* (cf. Exercise 2.2):

$$d_H(A, B) = \min_{\epsilon \geq 0} \{ A \subset N_\epsilon(B) \text{ and } B \subset N_\epsilon(A) \}.$$

This may not be so easy to see, but we are looking at each compact set as a point. The advantage of having a Hausdorff metric is clear in the following theorem.

THEOREM 2.1. *The Hausdorff metric d_H turns $K(\mathbf{R}^n)$ into a complete metric space.*

PROOF. Let $\{A_i\}$ be an arbitrary Cauchy sequence in $K(\mathbf{R}^n)$, so that for any number $\epsilon > 0$ we can select a large enough $m \equiv m(\epsilon)$ with $d_H(A_p, A_q) \leq \epsilon$ for every pair of integers p and q, $p \geq q \geq m$. We must show that the sequence $\{A_i\}$ converges with respect to the Hausdorff metric to some point in $K(\mathbf{R}^n)$.

Set

$$E_k = \overline{\bigcup_{i=k}^{\infty} A_i}.$$

Evidently the sequence $\{A_i\}$ is uniformly bounded so that each E_k is a compact set. Furthermore, since the sequence $\{E_k\}$ is monotone decreasing, the set

$$E = \bigcap_{k=1}^{\infty} E_k$$

belongs to $K(\mathbf{R}^n)$. Then it follows that

$$E \subset E_q = \overline{\bigcup_{i=q}^{\infty} A_i} \subset N_\epsilon(A_q).$$

On the other hand, given $x \in A_q$, there exists some point $y_p \in A_p$ such that $||x - y_p|| \leq \epsilon$, for every integer p, $p \geq q$, because $A_q \subset N_\epsilon(A_p)$. If z is an accumulation point of the sequence $\{y_p\}$, we have that $||x - z|| \leq \epsilon$. Moreover, since for any $p \geq k$ we have

$$y_p \in A_p \subset E_p \subset E_k,$$

and E_k is compact, we get $z \in E_k$, which in turn implies that $z \in E$ as k was arbitrary. Hence we have

$$x \in N_\epsilon(\{z\}) \subset N_\epsilon(E).$$

Thus $d_H(E, A_q) \leq \epsilon$; that is, the sequence $\{A_i\}$ converges to E. $\qquad\square$

We use this theorem to attain our goal in this section, which is to show the existence and uniqueness of self-similar sets.

THEOREM 2.2. *Given a family of* m *contractions* $\{\psi_1, \psi_2, \ldots, \psi_m\}$, $m \geq 2$, *there exists a unique self-similar set* V.

PROOF. We first define a map $\Phi : K(\mathbf{R}^n) \to K(\mathbf{R}^n)$ by

$$(2.3) \qquad \Phi(A) = \bigcup_{i=1}^{m} \psi_i(A),$$

where $\psi_i(A)$ on the right-hand side is the image of A by ψ_i. The well-known facts that the image of a compact set under a continuous map is compact and that a union of finitely many compact sets is compact assure us that Φ maps $K(\mathbf{R}^n)$ into $K(\mathbf{R}^n)$.

Since by Theorem 2.1 $K(\mathbf{R}^n)$ is a complete metric space, it remains to show that the map Φ is a contraction on $K(\mathbf{R}^n)$. Then its unique fixed point $F(\Phi) \in K(\mathbf{R}^n)$ becomes our desired self-similar set V.

Now for arbitrary sets $A_0, A_1, A_2, A_3 \in K(\mathbf{R}^n)$ the following properties hold:

(i) $d_H(\psi_i(A_0), \psi_i(A_1)) \leq L(\psi_i) d_H(A_0, A_1)$.
(ii) $d_H(A_0 \cup A_1, A_2 \cup A_3) \leq \max\{d_H(A_0, A_2), d_H(A_1, A_3)\}$.

To prove (i), put $s = d_H(A_0, A_1)$; then for any $x \in A_0$ there is some $y \in A_1$ such that $||x - y|| \leq s$. This implies that

$$||\psi_i(x) - \psi_i(y)|| \leq L(\psi_i)||x - y|| \leq L(\psi_i)s,$$

and hence $\psi_i(x) \in N_t(\psi_i(A_1))$, where we write for simplicity $L(\psi_i)s = t$. From this we get $\psi_i(A_0) \subset N_t(\psi_i(A_1))$, as we chose x arbitrarily. Similarly we get the second inclusion relation. Next, in order to prove (ii) we set $s = d_H(A_0, A_2)$ and $t = d_H(A_1, A_3)$; then the inclusions $A_0 \subset N_s(A_2)$ and $A_1 \subset N_t(A_3)$ give the inclusions

$$A_0 \cup A_1 \subset N_s(A_2) \cup N_t(A_3) \subset N_r(A_2 \cup A_3),$$

where $r = \max\{s, t\}$. The second inclusion comes about in the same way.

Now for any $A, B \in K(\mathbf{R}^n)$, the repeated use of the inequality (ii) results in

$$d_H(\Phi(A), \Phi(B)) = d_H\left(\bigcup_{i=1}^{m} \psi_i(A), \bigcup_{i=1}^{m} \psi_i(B)\right)$$
$$\leq \max_{1 \leq i \leq m} d_H\left(\psi_i(A), \psi_i(B)\right).$$

From the property (i), one gets

$$\text{RHS} \leq \left(\max_{1 \leq i \leq m} L(\psi_i)\right) d_H(A, B),$$

which, together with the obvious inequality

$$\max_{1 \leq i \leq m} L(\psi_i) < 1,$$

enables us to conclude that the map $\Phi : K(\mathbf{R}^n) \to K(\mathbf{R}^n)$ is a contraction.   □

The above result says that, starting with any compact set A, the sequence of compact sets $\{\Phi^k(A)\}$ converges to the self-similar set V. In particular, if a set B in $K(\mathbf{R}^n)$ satisfies $\Phi(B) \subset B$ (for example B can be a closed ball of a large enough radius centered at the origin), the sequence $\{\Phi^k(B)\}$ of compact sets is monotone decreasing and hence its limit $\bigcap \Phi^k(B)$ becomes the self-similar set V (the map Φ as a dynamical system has a globally stable fixed point $F(\Phi)$). In other words, just like the Cantor set C or the set S which we constructed in the first chapter, the intersection of a monotone decreasing sequence of compact sets represents the self-similar set V.

The same observation applies for $B \in K(\mathbf{R}^n)$ with $\Phi(B) \supset B$ (for example B could be the fixed point of ψ_i), and in this case we see that $V \supset B$.

2.2. The size and shape of a self-similar set

In order to measure the size of a self-similar set we first state some results concerning its Hausdorff dimension. As we saw in the previous chapter, evaluating the Hausdorff measure from above is easy.

Given a set of contractions $\{\psi_1, \psi_2, \ldots, \psi_m\}$, we define the *similarity dimension* of the corresponding self-similar set V to be the positive root of an equation in d

(2.4) $$\sum_{i=1}^{m} (L(\psi_i))^d = 1,$$

which depends only on the Lipschitz constants of the contractions. We denote this value by $\dim_S(V)$. Although we call $\dim_S(V)$ a dimension, it does not define a dimension function in the sense discussed in §1.1. We use the name because under a certain condition this value agrees with the Hausdorff dimension of the self-similar set V. We might mention that equation (2.4) has a unique positive root because the function $f(x) = \sum (L(\psi_i))^x$ is monotone decreasing and satisfies the relation $f(0+) = m \geq 2 > f(\infty) = 0$. We have in general the following inequality:

THEOREM 2.3. *For a self-similar set V,*

$$\dim_H(V) \leq \dim_S(V).$$

PROOF. It suffices to show $\mathcal{H}^s(V) < \infty$, where $s = \dim_S(V)$. Given a sub–set A of $\mathbf{R}^n$ and a sequence

$$i_1 i_2 \cdots i_k \qquad (1 \le i_j \le m, \ j = 1, 2, \ldots, k)$$

of k numbers ranging from 1 to m, we use the following abbreviation for the sake of simplicity:

$$A_{i_1 i_2 \cdots i_k} \equiv \psi_{i_1} \circ \psi_{i_2} \circ \cdots \circ \psi_{i_k}(A).$$

Then, using (2.2), we get

$$\bigcup_{i_1 i_2 \ldots i_k} V_{i_1 i_2 \cdots i_k} = V.$$

Furthermore, we have

$$|V_{i_1 i_2 \cdots i_k}| \le |V| \prod_{j=1}^{k} L(\psi_{i_j}) \le \lambda^k |V|,$$

where λ is the largest among the $L(\psi_i)$'s, which is of course less than one. We set $\delta = \lambda^k |V|$. Then the set $\{V_{i_1 i_2 \cdots i_k}\}$ is a δ-covering of V, and we have

$$\mathcal{H}^s_\delta(V) \le \sum_{i_1=1}^{m} \cdots \sum_{i_k=1}^{m} |V_{i_1 i_2 \cdots i_k}|^s$$

$$\le |V|^s \sum_{i_1=1}^{m} \cdots \sum_{i_k=1}^{m} \prod_{j=1}^{k} (L(\psi_{i_j}))^s$$

$$\le |V|^s \left(\sum_{i=1}^{m} (L(\psi_i))^s \right)^k = |V|^s.$$

Now we can downsize δ as much as we wish by taking k large enough. Thus we get the inequality $\mathcal{H}^s(V) \le |V|^s$. $\qquad\qquad \square$

Although the actual computation of the positive root d of (2.4) in general is hard, the value of d can be determined at least in theory, and it is just a question of computation to estimate its value up to some high degree of accuracy. In this sense it is very important to know when the inequality in Theorem 2.3 becomes an equality. If the equality holds then the Hausdorff dimension, which is difficult to determine in general, can be obtained by solving equation (2.4).

We say that the set of contractions $\{\psi_1, \psi_2, \ldots, \psi_m\}$ satisfies the *open set condition* if there exists a nonempty bounded open set $U \subset \mathbf{R}^n$ such that

$$\psi_i(U) \subset U, \quad 1 \le i \le m, \quad \text{and} \quad \psi_i(U) \cap \psi_j(U) = \emptyset, \quad i \ne j.$$

Since $\Phi(\overline{U}) \subset \overline{U}$, we have $V \subset \overline{U}$, as we mentioned at the end of §2.1. Further, we say that a map $\psi : \mathbf{R}^n \to \mathbf{R}^n$ is a *similar contraction* if $\|\psi(x) - \psi(y)\| = L(\psi)\|x - y\|$, for every x and every y in $\mathbf{R}^n$. In other words, the map ψ is a linear transformation expressed as a composition of similar contractions, rotations or inversions. We now have the following theorem of Hutchinson. We have modified the original proof to be more elementary without the concepts of symbol spaces and invariant measures.

THEOREM 2.4. *For a self-similar set V defined by a family of similar contractions which satisfies the open set condition, the equality $\dim_H(V) = \dim_S(V)$ holds.*

PROOF. Suppose that the bounded open set U in the open set condition contains a closed ball of radius α and is contained in a closed ball of radius β. Set

$$\gamma = \min_{1 \leq i \leq m} L(\psi_i) \in (0,1).$$

We show by contradiction that

$$\mathcal{H}^s(V) \geq \left(\frac{\alpha\gamma}{2\beta+1}\right)^n.$$

This inequality will then complete the proof, since $\mathcal{H}^s(V) \leq |V|^s$ by Theorem 2.3. In fact, we will have shown the stronger result that V is an s-set.

Now denote by τ the value of the RHS and suppose that $\mathcal{H}^s(V) < \tau$. Then for some small-enough positive number δ and a δ-covering $\{W_i\}$ of V the inequality

$$\sum_{i=1}^{\infty} |W_i|^s < \tau$$

holds. Here, as we can enlarge each W_i slightly while keeping the inequality, we may assume each W_i is an open set. The set V being compact, a finite number of the open sets in the family $\{W_i\}$ cover V. Hence, we may assume that we have a finite bounded open covering $\{W_i\}$, so that for some natural number N we have

(2.5) $$V \subset \bigcup_{i=1}^{N} W_i \quad \text{and} \quad \sum_{i=1}^{N} |W_i|^s < \tau.$$

For each $k \geq 1$ and an arbitrary set $A \in \mathbf{R}^n$, put

$$\mu_k(A) = \sum \prod_{j=1}^{k} (L(\psi_{i_j}))^s,$$

where the sum runs through every finite sequence $i_1 i_2 \cdots i_k$ which satisfies the condition $A \cap \overline{U}_{i_1 i_2 \cdots i_k} \neq \emptyset$. With A left fixed, the sequence $\{\mu_k(A)\}$ is monotone decreasing, because

(2.6) $$U_{i_1 i_2 \cdots i_k j} \subset U_{i_1 i_2 \cdots i_k}, \quad 1 \leq j \leq m, \quad \text{and} \quad \sum_{j=1}^{m} (L(\psi_j))^s = 1.$$

Thus the limit

$$\mu(A) = \lim_{k \to \infty} \mu_k(A)$$

exists. In particular, the inclusion $V \subset \overline{U}$ implies that $\mu(V) = 1$. We also have the following obvious properties:

(i) If $A \subset B$ then $\mu(A) \leq \mu(B)$.
(ii) $\mu(A \cup B) \leq \mu(A) + \mu(B)$.

Using these properties, we obtain the inequalities

$$1 = \mu(V) \leq \mu\left(\bigcup_{j=1}^{N} W_j\right) \leq \sum_{j=1}^{N} \mu(W_j).$$

We shall evaluate $\mu(W_j)$ next. As we may assume that $|W_j|$ is less than δ, there exists at least one natural number $k \geq 2$ for which the inequalities

$$L(\psi_{i_1}) \cdots L(\psi_{i_{k-1}}) > |W_j| \geq L(\psi_{i_1}) \cdots L(\psi_{i_k})$$

hold. Let Λ_j be the family of the sequences $\{i_1 \cdots i_k\}$ which satisfy the above relation. The lengths of the sequences in Λ_j are not uniform in general. Let l be the maximum value among these lengths. Then by (2.6), we have

$$\mu_l(W_j) \leq \sum (L(\psi_{i_1})) \cdots L(\psi_{i_k}))^s,$$

where the sum runs over all $i_1 \cdots i_k$ with $W_j \cap \overline{U}_{i_1 \cdots i_k} \neq \emptyset$. This is because the condition $W_j \cap \overline{U}_{i_1 \cdots i_k \cdots i_l} \neq \emptyset$ implies $W_j \cap \overline{U}_{i_1 \cdots i_k} \neq \emptyset$. It follows therefore that

$$\mu(W_j) \leq \mu_l(W_j) \leq p|W_j|^s,$$

where p is the number of sequences in Λ_j with $W_j \cap \overline{U}_{i_1 \cdots i_k} \neq \emptyset$.

Put

$$r = L(\psi_{i_1}) \cdots L(\psi_{i_k})\alpha \quad \text{and} \quad R = L(\psi_{i_1}) \cdots L(\psi_{i_k})\beta;$$

then as each ψ_i is a similar contraction, we see that $U_{i_1 \cdots i_k}$ contains a closed ball of radius r and is contained in a closed ball of radius R. If x is an arbitrary point of W_j the closed ball of radius $|W_j|$ centered at x contains W_j, and so $U_{i_1 \cdots i_k}$ is contained in some closed ball of radius $2R + |W_j|$. We also have that

$$r \geq \alpha\gamma L(\psi_{i_1}) \cdots L(\psi_{i_{k-1}}) > \alpha\gamma|W_j| \quad \text{and} \quad R \leq \beta|W_j|;$$

hence it turns out that $U_{i_1 \cdots i_k}$ sits in some closed ball of radius $(2\beta + 1)|W_j|$ and that it contains some closed ball of radius $\alpha\gamma|W_j|$. We remind the reader that the ball of radius $(2\beta + 1)|W_j|$ does not depend on the sequences in Λ_j. The $U_{i_1 \cdots i_k}$'s, on the other hand, are mutually disjoint, and so by comparing their n-dimensional volumes – the volume of a ball is proportionate to the n-th power of its radius – we get

$$p(\alpha\gamma|W_j|)^n \leq ((2\beta + 1)|W_j|)^n,$$

or equivalently,

$$p \leq \left(\frac{2\beta + 1}{\alpha\gamma}\right)^n = \tau^{-1}.$$

This implies that $\mu(W_j) \leq \tau^{-1}|W_j|^s$, which in turn gives:

$$1 \leq \sum_{j=1}^N \mu(W_j) \leq \tau^{-1} \sum_{j=1}^N |W_j|^s;$$

however, this contradicts (2.5). $\square$

REMARK. For the similar contractions $\{\psi_1, \psi_2, \ldots, \psi_m\}$ which satisfy the open set condition, the positive solution d of equation (2.4) – the similarity dimension of V – must be less than or equal to n. Thus we see that if a family of similar contractions has a similarity degree which is greater than n, then it does not meet the open set condition. Falconer[4], however, proved that the equation in Theorem (2.4) holds – except, here he replaces $\dim_S(V)$ by $\min\{\dim_S(V), n\}$ – in a large class of similar contractions which do not satisfy the open set condition.

We now ask the reader to compare Theorem 1.2 and Theorem 2.4. Why is it so difficult to determine $\mathcal{H}^s(V)$ for $n \geq 2$? The answer lies in the geometrical pattern of the coverings. In the one-dimensional case all we needed to consider was essentially one geometrical pattern, namely intervals, which are one-dimensional balls. Note that we also made a good use of balls in the proof of Theorem 2.4.

[4]Falconer, K. J., The Hausdorff dimension of self-affine fractals, Math. Proc. Cambridge Philos. Soc., **103**(1988), 339-350

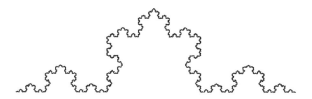

FIGURE 2.1. The Koch curve

For example, the Cantor set we defined in §1.3 is the sef-similar set corresponding to two similar contractions

$$\text{(2.7)} \qquad \psi_1(x) = \frac{x}{3}, \qquad \psi_2(x) = 1 - \frac{x}{3},$$

and as $L(\psi_1) = L(\psi_2) = 1/3$, equation (2.4) defining the similarity dimension becomes

$$\left(\frac{1}{3}\right)^d + \left(\frac{1}{3}\right)^d = 1.$$

Hence, we get $\dim_S(C) = \log 2 / \log 3$. In this case, the open set condition is satisfied with $U = (0, 1)$. If we replace $\psi_2(x) = 1 - x/3$ by $\psi_2(x) = (x + 2)/3$, we still get the same Cantor set.

We have thus seen that different sets of similar contractions may correspond to the same self-similar set. Now the set S is the self-similar set for three similarity contractions of $\mathbf{R}^2$, each with the Lipschitz constant $1/3$, and by solving

$$\left(\frac{1}{3}\right)^d + \left(\frac{1}{3}\right)^d + \left(\frac{1}{3}\right)^d = 1$$

we find its similarity dimension to be one: $\dim_S(S) = 1$. In this case these contractions satisfy the open set condition with $U = (0, 1) \times (0, 1)$. These results in both cases agree with what we saw in Chapter One. We note also that they are both totally disconnected. The self-similar set K corresponding to the similar contractions of $\mathbf{C} \approx \mathbf{R}^2$:

$$\text{(2.8)} \qquad \psi_1(z) = \omega \bar{z}, \qquad \psi_2(z) = \bar{\omega}(\bar{z} - 1) + 1,$$

where $\omega = 1/2 + \sqrt{3}i/6$, is a continuous Jordan curve, called the *Koch curve*, which admits no tangent line anywhere (Figure 2.1).

Since $L(\psi_1) = L(\psi_2) = |\omega| = 1/\sqrt{3}$, by solving

$$\left(\frac{1}{\sqrt{3}}\right)^d + \left(\frac{1}{\sqrt{3}}\right)^d = 1,$$

we get $\dim_S(K) = \log 4 / \log 3$. This is exactly twice the dimension of the Cantor set. The contractions in this case satisfy the open set condition if we take the interior of the isosceles triangle with vertices at $0, 1$ and ω for U. As the topological dimension of K is one, we conclude that K also is a fractal.

We take up, next, the connectivity of self-similar sets, the most basic question with regard to their geometric shapes. Here we give a necessary and sufficient condition for a self-similar set to be connected. We first need some definitions.

We say that m sets $\{A_1, A_2, \ldots, A_m\}$ in $\mathbf{R}^n$ form a chain when for any $k \neq j$, passing to a sequence $i_1 \cdots i_l$ if necessary, each of the sets

$$A_k \cap A_{i_1}, \ A_{i_1} \cap A_{i_2}, \cdots, A_{i_{l-1}} \cap A_{i_l}, \ A_{i_l} \cap A_j$$

is non-empty. We say that the set A forms an ϵ-chain when for an arbitrary pair of points $x, y \in A$, passing to a finite sequence of points $\{z_1, z_2, \ldots, z_k\} \subset A$ if necessary, each of the values

$$\|x - z_1\|, \ \|z_1 - z_2\|, \ldots, \|z_{k-1} - z_k\|, \ \|z_k - y\|$$

is less than or equal to ϵ. Then the following two properties are evident:

(a) If each A_i forms an ϵ-chain and if the family $\{A_1, A_2, \ldots, A_m\}$ forms a chain, then $\bigcup_{i=1}^m A_i$ also forms an ϵ-chain.

(b) If A forms an ϵ-chain, then for any contraction $\psi : \mathbf{R}^n \to \mathbf{R}^n$, $\psi(A)$ forms an $L(\psi)\epsilon$-chain.

We now have

THEOREM 2.5. *The self-similar set V is connected if and only if the set family $\{\psi_1(V), \psi_2(V), \ldots, \psi_m(V)\}$ forms a chain.*

PROOF. If V is connected, then the family of sets $\{\psi_1(V), \ldots, \psi_m(V)\}$ clearly forms a chain. We show the converse by contradiction.

Suppose that V is not connected, so that it gets decomposed into two non-empty compact sets A_1 and A_2 (*i. e.*, $V = A_1 \cup A_2$, $A_1 \cap A_2 = \emptyset$). We then see that

$$\delta = \min_{x \in A_1, y \in A_2} \|x - y\| > 0,$$

and that V never forms an ϵ-chain for any ϵ smaller than δ. Meanwhile, if we denote by λ the largest number among the $L(\psi_i)$, V evidently forms a $|V|$-chain and so, by the property (b), each $V_i = \psi_i(V)$ forms a $\lambda|V|$-chain. Then according to the property (a), the set $\bigcup V_i$ also forms a $\lambda|V|$-chain, which is equal to V by (2.2) for self-similar sets. We can repeatedly use the above argument to conclude that V forms a $\lambda^l|V|$-chain for any natural number l. Since $\lambda < 1$, if l is sufficiently large we get $\lambda^l|V| < \delta$. But this is a contradiction. □

The following *Williams's formula* gives us an effective tool in deciding when the family $\{\psi_1(V), \ldots, \psi_m(V)\}$ forms a chain:

$$V = \overline{\bigcup_{\substack{i_1 \cdots i_k \\ k \geq 1}} F(\psi_{i_1} \circ \cdots \circ \psi_{i_k})},$$

where the union is taken over all finite sequences $i_1 \cdots i_k$, $k \geq 1$. This is the formula for the solution of equation (2.2) of the self-similar set V. We see, in particular, that V contains the point

$$\psi_{i_1} \circ \cdots \circ \psi_{i_k}(F(\psi_{j_1} \circ \cdots \circ \psi_{j_l}))$$

for an arbitrary pair of finite sequences $i_1 \cdots i_k$ and $j_1 \cdots j_l$, which for simplicity we shall denote by $[i_1 \cdots i_k; j_1 \cdots j_l]$. We can actually calculate this point without knowing V; therefore, if it belongs to the intersection of V_i and V_j, then by Theorem 2.5 the connectivity of V will follow.

For example, for the Koch curve in Figure 2.1 we have $\psi_1(F(\psi_2)) = \omega = \psi_2(F(\psi_1))$ (check this); that is, $[1;2] = [2;1]$, and $\omega \in K_1$ and $\omega \in K_2$. Hence the pair $\{K_1, K_2\}$ forms a chain, and so K is connected.

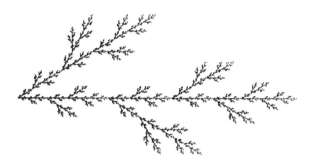

FIGURE 2.2. The tree

Another example is the self-similar set called a *tree* (Figure 2.2) corresponding to the following two similar contractions of $\mathbf{C}$:

$$\psi_1(z) = \omega\bar{z}, \qquad \psi_2(z) = \frac{2\bar{z}+1}{3}.$$

Unlike the Koch curve, this set has infinitely many endpoints. Since $L(\psi_1) = 1/\sqrt{3}$ and $L(\psi_2) = 2/3$, the similarity dimension is the (positive) root of the equation

$$\left(\frac{1}{\sqrt{3}}\right)^d + \left(\frac{2}{3}\right)^d = 1.$$

In the above example, if we take the interior of the pentagon with vertices at 0, $\bar{\omega}$, 1, ω, ω^2 for the set U, we see that the defining contractions satisfy the open set condition. Hence, the Hausdorff dimension is also equal to the positive root of the equation above, and we see that the tree is fractal (since $1 < d < 2$). Moreover, the tree is connected, since the equality $[2;1] = 1/3 = [11;2]$ implies that $1/3 \in V_1$ and $1/3 \in V_2$, and this in turn says that the pair $\{V_1, V_2\}$ forms a chain.

2.3. Self-affine sets

We call a self-similar set defined by linear contractions a *self-affine set*. In this section we study self-affine sets determined by two linear transformations of the plane ($n = m = 2$), the most basic case in our present situation. We express such a transformation as a transformation of $\mathbf{C}$, using complex numbers, by

$$\psi(z) = az + b\bar{z} + c,$$

where without loss of generality we may normalize ψ so that $z = 0$ is the fixed point of ψ_1 and $z = 1$ is the fixed point of ψ_2. We then obtain the following normal form:

$$\begin{pmatrix} \psi_1(z) \\ \psi_2(z) \end{pmatrix} = \begin{pmatrix} \alpha & \beta \\ \gamma & \delta \end{pmatrix} \begin{pmatrix} z \\ \bar{z} \end{pmatrix} + \begin{pmatrix} 0 \\ 1 - \gamma - \delta \end{pmatrix},$$

where $\{\alpha, \beta, \gamma, \delta\}$ are complex parameters. We call the matrix $\begin{pmatrix} \alpha & \beta \\ \gamma & \delta \end{pmatrix}$ a parameter matrix. We may assign the self-similar set V defined by the pair $\{\psi_1, \psi_2\}$ to this parameter matrix. Here naturally we impose the following condition so as to make each ψ_i a contraction:

$$0 < |\alpha| + |\beta| < 1 \quad \text{and} \quad 0 < |\gamma| + |\delta| < 1,$$

the reason being that with respect to ψ_1, for example, we have

$$\|\psi_1(z) - \psi_1(w)\| = \|\alpha(z - w) + \beta(\bar{z} - \bar{w})\|$$
$$\leq (|\alpha| + |\beta|)\|z - w\|.$$

Note further that each ψ_i is a similar contraction only when $\alpha\beta = \gamma\delta = 0$ (only when one factor in each product is zero).

It is a known fact that totally disconnected non-trivial self-affine sets such as the set S cited in §1.3 are all homeomorphic to the Cantor set C. Thus, if one's interest lies in geometric shapes, one might as well deal only with connected affine sets. Generally speaking, however, it is difficult to decide when a parameter matrix defines a connected V, and as we mentioned after Theorem 2.5, it becomes necessary to locate a point of $V_1 \cap V_2$ in some clever way. This is where we use the connectivity condition relating the respective fixed points, called the *basic relation*, which we write as a rule:

$$[1i_1 \cdots i_k; j_1 \cdots j_l] = [2p_1 \cdots p_s; q_1 \cdots q_r].$$

We suspect that this basic relation indeed reflects the geometrical nature of the self-affine set, so that it will be one yardstick for the classification of self-affine sets.

We have, for example, the following Table 2.1 which lists the self-affine sets satisfying the simplest basic relation $[1; 2] = [2; 1]$. Here $\sigma = (1+i)/2$, $\tau = (2+i)/4$.

Table 2.1. Self-affine sets

parameters α β γ δ	1/2 $\quad$ 0 $\quad$ 0 $\quad$ 1/2	0 $\quad$ ω $\quad$ 0 $\quad$ $\bar{\omega}$	0 $\quad$ σ $\quad$ 0 $\quad$ $\bar{\sigma}$	σ $\quad$ 0 $\quad$ $\bar{\sigma}$ $\quad$ 0	σ $\quad$ 0 $\quad$ 0 $\quad$ $\bar{\sigma}$	τ $\quad$ $i/4$ $\quad$ $\bar{\tau}$ $\quad$ $-i/4$
self-affine set	the interval $[0, 1]$	the Koch curve	right isosceles triangle (Pólya's curve)	the Lévy curve	the procession of crabs	the Takagi graph

All self-affine sets are single curves, among which some are self-intersecting. There are even some with the positive area. Pólya's curve – a simplified version of the Peano curve – has the area of a right isosceles triangle. One gets Lévy's curve (Figure 2.3) by interchanging the columns of the parameter matrix of the Pólya curve. Although Lévy's curve looks quite porous, just as a curve should, in reality its area equals that of the right isosceles triangle with vertices at 0, 1 and σ.

The Pólya curve satisfies the open set condition, as one can take the interior of the triangle for U. Of course each of these curves has a positive area, and so its Hausdorff dimension is two.

The procession of the crabs (Figure 2.4) has a parameter matrix resembling the above two, but its configuration is completely different. We lastly mention that the graph of the Takagi function (Figure 1.7) is also self-affine, justifying our statement in §1.4 that this graph is self-similar. The basic relation $[1; 2] = [2; 1]$, which is common to all the examples given here, is expressed in terms of the matrix entries:

$$\alpha + \beta + \gamma + \delta = 1.$$

More examples. We look at the basic relation $[12; 1] = [21; 2]$:

FIGURE 2.3. Lévy's curve

FIGURE 2.4. The procession of the crabs

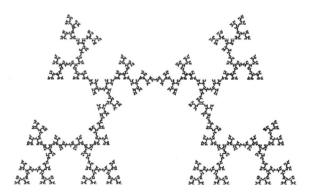

FIGURE 2.5. The result of scooping out regular pentagons *ad infinitum*

FIGURE 2.6. The set D

parameter matrix $\begin{matrix} \alpha & \beta \\ \gamma & \delta \end{matrix}$	$\begin{matrix} 0 & \xi \\ 0 & \bar{\xi} \end{matrix}$	$\begin{matrix} \sigma & 0 \\ \sigma & 0 \end{matrix}$
self-affine set	Fig.2.5	Fig. 2.6

Here $\xi = 1 - e^{-\pi i/5}$. We get the configuration in Figure 2.5 by scooping regular pentagons out repeatedly *ad infinitum* from the irregular pentagon whose vertices are 0, 1, $1 - \bar{\xi}$, $\xi(1 - \xi)$, ξ. The remaining set satisfies the open set condition; to see this, take the interior of the irregular pentagon as the open set U.

The set D in Figure 2.6 is given by the parameter matrix obtained from Lévy's curve where we replace $\bar{\sigma}$ by σ. The area of D also is positive. The set D shares with squares and regular hexagons the distinguishing feature that its parallel translations fill the plane; we can express this fact by

$$\bigcup_{p,q \in \mathbf{Z}} (D + p + qi) = \mathbf{C},$$

where the set $D + w$ stands for the set $\{z + w; z \in D\}$.

2.4. Fractals and chaos

In §2.2 we defined by the equations (2.7) a couple of similar contractions $\{\psi_1, \psi_2\}$ on $\mathbf{R}^1$ whose self-similar set turned out to be the Cantor set. On the other hand, given the piecewise linear function on $\mathbf{R}^1$ defined by

$$f(x) = \frac{3}{2} - \left| 3x - \frac{3}{2} \right|,$$

starting with the initial value $x_0 \in \mathbf{R}$ we can recursively determine the sequence of points $\{x_k\}$ by

$$x_{k+1} = f(x_k).$$

In this way the function $f(x)$ defines a one-dimensional *discrete dynamical system*, and we say that the sequence $\{x_k\}$ is the *orbit* starting at x_0.

For instance, the orbit which starts at $x_0 = 1/2$ diverges to $-\infty$. There are, however, initial values for which their orbits do not diverge. Among these, for example, we see that $x_0 = 0$ and $x_0 = 3/4$ are the fixed points of $f(x)$. More generally, the k-periodic points of $f(x)$ also give orbits which do not diverge. What then is the set of initial values for which the orbits do not diverge? The answer is the Cantor set C. In fact the ψ_i's defined by (2.7) are the two inverse functions of the dynamical system $f(x)$ under consideration, and from the formula for the self-similar set

$$\psi_1(C) \cup \psi_2(C) = C,$$

it follows $f(C) = C$. The orbit starting at a point $x \in C$ will stay forever in C. We say that C is the *invariant set* of the dynamical system $f(x)$.

Consider the intervals $I_1 = [0, 1/3]$ and $I_2 = [2/3, 1]$, and an arbitrary infinite sequence $i_1 i_1 i_2 \cdots$ consisting only of 1 or 2. Then we can always find an initial value x_0 such that $x_k \in I_{i_k}$ holds for every $k \geq 0$. In this sense, we say that the dynamical system $f(x)$ is *chaotic*. In fact, we may set

$$x_0 = \bigcap_{k=0}^{\infty} \psi_{i_0} \circ \cdots \circ \psi_{i_k}([0,1]).$$

In other words, we can build an orbit for a randomly chosen sequence; what this means is that there are infintely many types of orbits available, making the dynamical system $f(x)$ indeed very complicated.

We have seen above that the two contractions corresponding to the Cantor set are the inverse functions of the chaotic discrete dynamical system $f(x)$, and so in this sense we might say that fractals and chaos are inversely related. Not every self-similar set corresponds inversely to a chaotic dynamical sytem; however, at least for any totally disconnected self-similar set we can construct an inversely related chaotic dynamical system just as we did for the Cantor set. Sometimes even if the set V is connected we can define an inverse chaos. The Koch curve K defined by the equation (2.8), for instance, corresponds to the following discrete dynamical system on $\mathbf{C}$:

$$g(z) = \begin{cases} \bar{z}/\bar{\omega}, & \Re z \geq 1/2, \\ (1-z)/\bar{\omega}, & \Re z < 1/2, \end{cases}$$

and K is the invariant set of $g(z)$. When $\Re z = 1/2$ we have $z + \bar{z} = 1$; thus $g(z)$ is continuous (our Koch curve here is the self-similar set defined by $\psi_1(z) = \omega \bar{z}$ and $\psi_2(z) = 1 - \bar{\omega} z$).

Exercises

2.1. (*The contraction principle for a complete metric space*) Let Y be a complete metric space and $\psi : Y \to Y$ a contraction. Show that ψ has exactly one fixed point in Y. [Hint: Starting with an arbitrary point $y_0 \in Y$, define a sequence $\{y_n\}$ by $y_{n+1} = \psi(y_n)$, and show that this is Cauchy.]

2.2. Show that the Hausdorff metric d_H defined in §2.2 indeed satisfies the following metric axioms. Here A, B, and C are arbitrary compact sets.

(1) $d_H(A, B) = 0$ if and only if $A = B$.
(2) $d_H(A, B) = d_H(B, A)$.
(3) $d_H(A, B) \leq d_H(A, C) + d_H(C, B)$.

2.3. Given a set of contractions $\{\psi_1, \psi_2, \ldots, \psi_m\}$, let $\Phi : K(\mathbf{R}^n) \to K(\mathbf{R}^n)$ be the contraction defined by equation (2.3). Show that for a sufficiently large closed ball B we have $\Phi(B) \subset B$.

2.4. Show that if there exists a compact set A such that $\Phi(A) \supset A$, then $V \supset A$ for the corresponding self-similar set V.

2.5. For the contractions $\{\psi_1, \psi_2\}$ defined by (2.7), find an open set U, other than the open interval $(0, 1)$, which works for the open set condition.

2.6. Suppose that a set of similar contractions of $\mathbf{R}^n$ satisfies $\dim_S(V) = n$. Show that if the contractions satisfy the open set condition with an open set U, then $V = \overline{U}$. In particular, U is uniquely determined (cf. Exercise 2.5).

2.7. Find the basic relation for the self-affine set corresponding to each of the following parameter matrices.

$$(1) \begin{pmatrix} \sigma & 0 \\ 0 & -\sigma \end{pmatrix}, \qquad (2) \begin{pmatrix} 0 & \sigma \\ 0 & -\bar{\sigma} \end{pmatrix}, \qquad (3) \begin{pmatrix} 0 & \sigma \\ -\sigma & 0 \end{pmatrix},$$

where $\sigma = (1 + i)/2$.

2.8. Use the fact that the graph of the Takagi function is a self-affine set to derive the functional equation satisfied by $T(x)$ in such a way that $T(x)$ is the unique continuous solution of this equation.

2.9. Find a set of contractions $\{\psi_1, \psi_2\}$ (determining a self-affine set) that has at least two basic relations.

2.10. Consider a one-dimensional discrete dynamical system $f(x)$ corresponding to the contractions which define the Cantor set (cf. §2.4). Show that for any point x of C with

$$x = \sum_{n=1}^{\infty} \frac{a_n}{3^n}, \qquad a_n \in \{0, 2\},$$

we have

$$f(x) = \sum_{n=1}^{\infty} \frac{|a_1 - a_{n+1}|}{3^n}.$$

CHAPTER 3

An Alternative Computation for Differentiation

The main purpose of this chapter is to show that when we broaden the idea of differential equations, the Takagi function which we discussed in the first chapter becomes a unique solution of a certain boundary value problem. From this we will see that the Takagi function is related to yet another special function through a rather simple relation.

To obtain this result we consider again a certain chaotic dynamical system. The Takagi function turns out to be a generating function for this system. This fact, together with the Schauder expansion of a certain classical continuous function, leads us with ease to infinitely many systems of difference equations that the Takagi function satisfies. Moreover, we can do the same for Lebesgue's singular function, and we will find that these two functions are related in a simple way. This chapter also includes a brief discussion of the wavelet expansion, which resembles the Schauder expansion.

3.1. A chaotic dynamical system and its generating function

We consider the discrete dynamical system

$$(3.1) \qquad x_{n+1} = ax_n(1 - x_n), \qquad 0 \leq a \leq 4.$$

The case $a = 4$ is particularly interesting, where the system becomes the most chaotic. In this case, we can write the solution x_n of the initial value problem $x_{n+1} = 4x_n(1 - x_n)$ as an elementary function of n and the initial value x_0:

$$(3.2) \qquad x_n = \sin^2(2^n \arcsin \sqrt{x_0}), \qquad 0 \leq x_0 \leq 1.$$

We see immediately, using the trigonometric formula, that this equation satisfies equation (3.1). Furthermore, as we saw in Chapter Two, the system (3.1) is chaotic (here, $I_1 = [0, 1/2]$ and $I_2 = [1/2, 1]$).

On the other hand, we mentioned in §1.4 that Weierstrass studied the following function in 1872:

$$(3.3) \qquad W_{a,b}(x) = \sum_{n=0}^{\infty} a^n \cos(b^n \pi x),$$

where $0 < a < 1$, b is odd, and $ab > 1 + 3\pi/2$. This is the famous *continuous nowhere differentiable function* of Weierstrass. In 1916, Hardy weakened the condition on b, only requiring that b be a positive number with $ab \geq 1$, and obtained the similar but slightly weaker result that $W_{a,b}(x)$ *has finite derivatives nowhere* (Figure 1.5 shows the special case $b = 1/a = 2$).

We look therefore at the case $a = 1/2$, $b = 2$: even here the function defined by (3.3) is continuous with finite derivatives nowhere. Specifically, we have the function $W_{\frac{1}{2},2}$ defined by

(3.4) $$W_{\frac{1}{2},2}(x) = \sum_{n=0}^{\infty} \left(\frac{1}{2}\right)^n \cos(2^n \pi x).$$

Now it becomes evident how the above function relates to (3.2). Set $\Psi(x) = 4x(1-x)$. Then we get

(3.5) $$\sum_{n=0}^{\infty} \left(\frac{1}{2}\right)^n \Psi^n(x) = 1 - \frac{1}{2}W_{\frac{1}{2},2}\left(\frac{2}{\pi}\arcsin\sqrt{x}\right),$$

which, with $a = t$, generalizes to

(3.6) $$\sum_{n=0}^{\infty} t^n \Psi^n(x) = \frac{1}{2(1-t)} - \frac{1}{2}W_{t,2}\left(\frac{2}{\pi}\arcsin\sqrt{x}\right),$$

where $0 < t < 1$. We call the LHS of (3.6) the generating function of the chaotic dynamical system

(3.7) $$x_{n+1} = 4x_n(1-x_n).$$

We may say that *Weierstrass's function with the change of variable* $(2/\pi)\arcsin\sqrt{x}$ *differs from the generating function of the dynamical system* (3.2) *by* $1/2(1-t)$. Thus we may want to consider a more general case of this type of a function.

Let $g(x)$ be a bounded measurable function defined on the interval $[0,1]$ and let $\Psi(x)$ be a discrete dynamical system regarded as a map from $[0,1]$ to itself. Then we have the function $F(t,x)$ of two variables defined by

(3.8) $$F(t,x) = \sum_{n=0}^{\infty} t^n g(\Psi^n(x)),$$

which, for instance, defines Weierstrass's function if $\Psi(x) = 4x(1-x)$ and $g(x) = x$. We define the function $\varphi(x)$ by

(3.9) $$\varphi(x) = \begin{cases} 2x, & 0 \le x \le 1/2, \\ 2(1-x), & 1/2 \le x \le 1. \end{cases}$$

Now consider again the chaotic dynamical system $x_{n+1} = \varphi(x_n)$. With $t = 1/2$ and $g(x) = \varphi(x)/2$, the Takagi function $T(x)$ emerges as the generating function of (3.9) (see Figure 1.6), and for the same $\varphi(x)$ but with $g(x) = \cos \pi x$, Weierstrass's function comes up.

The function $F(t,x)$ satisfies the following functional equation:

(3.10) $$F(t,x) = tF(t,\Psi(x)) + g(x), \qquad |t| < 1, \quad x \in [0,1].$$

Conversely, given $g(x)$ and $\Phi(x)$, we find that the function $F(t,x)$ defined by (3.8) is the unique bounded solution which satisfies $F(0,x) = g(x)$.

We can easily solve, using the functional equation (3.10), the following

Inverse Problem. Let f be a function defined on $[0,1]$ and let Ψ be a dynamical system on $[0,1]$. Define $g(x)$ by $g_s(x) = f(x) - sf(\Psi(x))$. Then $F(s,x) = f(x)$, where $F(t,x)$ is a unique function which satisfies (3.10).

EXAMPLE 3.1. Let $f(x) = 2x - x^2$ and $\Psi(x) = \varphi(x)$. Set

$$g_s(x) = (2 - 4s)x + (4s - 1)x^2.$$

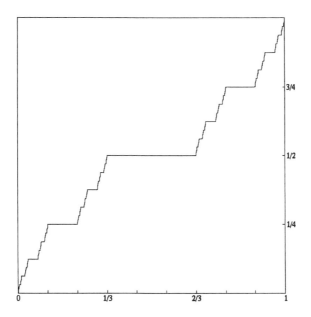

FIGURE 3.1. The Cantor function

Then we have the equality

$$(2 - 4s) \sum_{n=0}^{\infty} s^n \varphi^n(x) + (4s - 1) \sum_{n=0}^{\infty} s^n (\varphi^n(x))^2 = 2x - x^2;$$

in particular, for $s = 1/4$, this becomes

$$\sum_{n=0}^{\infty} \frac{1}{4^n} \varphi^n(x) = \sum_{n=0}^{\infty} \frac{1}{2^n} (\varphi^n(x))^2 = 2x - x^2, \qquad 0 \le x \le 1.$$

With a slight modification we get

$$(3.11) \qquad \sum_{n=1}^{\infty} \frac{1}{4^n} \varphi^n(x) = x - x^2 = x(1 - x),$$

which is the series expansion by φ of the polynomial $x - x^2$. If you compare this expression with the definition of the Takagi function

$$(3.12) \qquad \sum_{n=1}^{\infty} \frac{1}{2^n} \varphi^n(x) = T(x),$$

you will see that the LHS's of (3.11) and (3.12) differ respectively only by 2 and 4. This (3.12) is the continuous function (Fig. 1.7) which Takagi announced in 1903 (cf. Exercise 3.1).

REMARK. The equation (3.11) dates back to the time of Archimedes, *circa* 200 A. D., who used it as a quadratic *approximation by squeezing* to calculate the area bounded by a parabola and an arc. Note that it is different from later approximations such as Newton's, which use regular broken lines.

EXAMPLE 3.2. (The Cantor set and the Cantor funcion). Define $\Psi(x)$ by

$$\Psi(x) = \begin{cases} 3x, & 0 \leq x \leq 1/3, \\ 0, & 1/3 < x \leq 2/3, \\ 3x - 2, & 2/3 \leq x \leq x. \end{cases}$$

The set $\bigcap_{n=0}^{\infty} \Psi^{-n}([0,1])$ turns out to be Cantor's ternary set which we discussed in §1.3 (Fig. 1.2). Hence if we use $\frac{1}{2}\chi_{(\frac{1}{3},1]}(x)$ as the function $g_{\frac{1}{2}}(x)$ of the inverse problem, we find for $f(x)$ the *Cantor function* (Fig. 3.1), which is non-decreasing in the interval $[0,1]$ and has zero derivative almost everywhere. Here $\chi_{(\frac{1}{3},1]}(x)$ is the characteristic function taking the value one in $(1/3,1]$ and identically zero elsewhere.

The highlight of this section is the broken-line approximations of functions such as those defined in (3.11) and (3.12). Unlike Newton's approach, these are the approximations by similar triangles of varying scales.

3.2. The Schauder expansion

The reader must be familier with the Fourier expansion of a function, which expands the function as a linear combination of infinitely many trigonometric functions $\sin(nx)$ and $\cos(nx)$ $(n = 1, 2, \dots)$ and a constant number. The coefficients of the expansion are called Fourier coefficients. The Fourier expansion transforms a function of x into a periodic function of period n.

The Schauder expansion, on the other hand, has the following functions for its basis of expansion instead of $\sin(nx)$ and $\cos(nx)$. We define first a two-parameter function $F_{\alpha,\beta}(x)$ by

$$(3.13) \quad F_{\alpha,\beta}(x) = \begin{cases} \dfrac{1}{\beta - \alpha}\{|x - \alpha| + |x - \beta| - |2x - \alpha - \beta|\}, & \alpha \leq x \leq \beta, \\ 0, & x < \alpha, \ x > \beta. \end{cases}$$

The graph of the function $F_{\alpha,\beta}$ is the regular isosceles triangle of unit height based at α and β. The basis of the Schauder expansion consists of the following:

$$(3.14) \quad \begin{array}{c} 1, x, F_{0,1}, F_{0,\frac{1}{2}}, F_{\frac{1}{2},1}, F_{0,\frac{1}{4}}, F_{\frac{1}{4},\frac{1}{2}}, F_{\frac{1}{2},\frac{3}{4}}, F_{\frac{3}{4},1}, \dots, \\ \dots, F_{\frac{i}{2^k}, \frac{i+1}{2^k}}, \dots. \end{array}$$

Here the parameters α and β are $i/2^k$ and $(i+1)/2^k$ respectively, and $F_{\alpha,\beta}$ is defined for every natural number k and every natural number i $(0 \leq i \leq 2^k - 1)$.

THEOREM 3.1. *A continuous function f defined on the interval $[0,1]$ has a series expansion with the basis* (3.14), *which converges uniformly:*

$$(3.15) \quad f(x) = a_0 + a_1 x + \sum_{i,k} a_{k,i} F_{\frac{i}{2^k}, \frac{i+1}{2^k}}(x),$$

where $a_{k,i}$ is a number uniquely determined by f:

$$(3.16) \quad a_{k,i} = f\left(\frac{2i+1}{2^{k+1}}\right) - \frac{1}{2}\left\{f\left(\frac{i}{2^k}\right) + f\left(\frac{i+1}{2^k}\right)\right\}.$$

PROOF. This approximation is familiar and easy to understand. We show the first few terms in Figure 3.2.

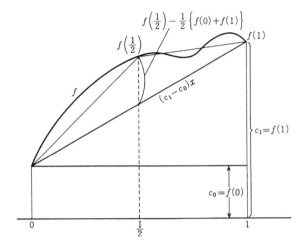

FIGURE 3.2. A broken-line approximation

The figure suggests that it is enough to show that the partial sum of n terms of the series is equal to the broken-line function tangent to the graph of f. To do this, we consider the $(n + 1)$-st partial sum:

$$S_n(x) = f(0) + \{f(1) - f(0)\}x$$
$$+ \sum_{k=0}^{n-1} \sum_{i=1}^{2k-1} a_{k,i} F_{\frac{i}{2^k}, \frac{i+1}{2^k}}(x).$$

Thus we only need to show that $S_n(i/2^k)$ and $f(i/2^k)$ agree at the corner point $i/2^k$ for each i, $0 \le i \le 2^k - 1$. But we can prove this by induction. $\square$

Let $\chi_{[a,b]}(x)$ be the characteristic function of the interval $[a, b]$. We have the following

Corollary. *Let f be a continuous function on $[0, 1]$. Then it has the expansion*

$$(3.17) \qquad fx) = f(0) + \{f(1) - f(0)\}x + \sum_{k=0}^{\infty} b_k(x)\varphi^{k+1}(x),$$

where

$$(3.18) \qquad b_k(x) = \sum_{i=0}^{2^k-1} a_{k,i}\chi_{[\frac{i}{2^k}, \frac{i+1}{2^k}]}(x).$$

The system of difference equations. We now apply (3.17) to the Takagi function defined by (3.12). Here we have $b_k(x) \equiv i/(2^{k+1})$. We see from (3.16) that the function $T(x)$ satisfies the infinite system of difference equations:

$$(3.19) \qquad T\left(\frac{2i+1}{2^{k+1}}\right) - \frac{1}{2}\left\{T\left(\frac{i}{2^k}\right) + T\left(\frac{i+1}{2^k}\right)\right\} = \frac{1}{2^{k+1}},$$
$$0 \le i \le 2^k - 1, \ k = 0, 1, 2, \ldots,$$

where the boundary values are $T(0) = T(1) = 0$.

We now look at the function f defined by (3.11) for comparison. In this case, we have $f(x) = x(1-x)$, and so

$$f\left(\frac{2i+1}{2^{k+1}}\right) - \frac{1}{2}\left\{f\left(\frac{i}{2^k}\right) + f\left(\frac{i+1}{2^k}\right)\right\} = \frac{1}{4^{k+1}},$$

(3.20)
$$0 \le i \le 2^k - 1, \ k = 0, 1, 2, \ldots,$$

$$f(0) = f(1) = 0.$$

The formulae (3.20) and (3.19) differ only by 4 and 2 on their respective RHS's. Moreover, the substitution $2^{-(k+1)} = h$, $(2i+1)/2^{k+1} = x$ in (3.20) gives us

$$f(x) - \frac{1}{2}\{f(x-h) + f(x+h)\} = h^2,$$

(3.21)
$$f(x+h) - f(x) - \{f(x) - f(x-h)\} = -2h^2,$$

$$\frac{f(x+h) - f(x) - \{f(x) - f(x-h)\}}{h^2} = -2.$$

The LHS of the last equation is the difference for d^2f/dx^2.

We might thus conclude that the formula (3.20) expresses the difference of the usual Dirichlet problem:

(∗)
$$\frac{d^2 f}{dx^2} = -2, \quad f(0) = f(1) = 0.$$

The function $f(x)$ is certainly a solution of (∗). We can deduce from this that *the Takagi function $T(x)$ is a unique solution of the discrete Dirichlet problem* (3.19). We can show that the solution is unique at the points $i/2^k$ by the formula (3.19) with the boundary values. Then the uniqueness of the solution follows from the fact that the set of binary rational points is dense in $[0, 1]$.

3.3. The de Rham equations and Lebesgue's singular function

Let us write the functional equation discussed in §3.1 for the function $T(x)$:

$$T(x) = \frac{1}{2}T(\varphi(x)) + \frac{\varphi(x)}{2}.$$

Using $\varphi(x)$ defined by (3.9), we get

$$\begin{cases} T(x) = \frac{1}{2}T(2x) + x, & 0 \le x \le \frac{1}{2}, \\ T(x) = \frac{1}{2}T(2(1-x)) + 1 - x, & \frac{1}{2} \le x \le 1, \end{cases}$$

which we rewrite as

(3.22)
$$\begin{cases} T(x) = \frac{1}{2}T(2x) + x, & 0 \le x \le \frac{1}{2}, \\ T(x) = \frac{1}{2}T(2x-1) + 1 - x, & \frac{1}{2} \le x \le 1. \end{cases}$$

We now recall the work of de Rham from 1957[1], which we had until quite recently thought irrelevant to our topic. He considered the following equations, very much like (3.22):

(3.23)
$$\begin{cases} M(x) = \alpha M(2x), & 0 \le x \le \frac{1}{2}, \\ M(x) = (1-\alpha)M(2x-1) + \alpha, & \frac{1}{2} \le x \le 1, \end{cases}$$

---

[1]de Rham, G., Sur un exemple de fonction continue sans dérivée, Enseign. Math., 3(1957), 71–72.

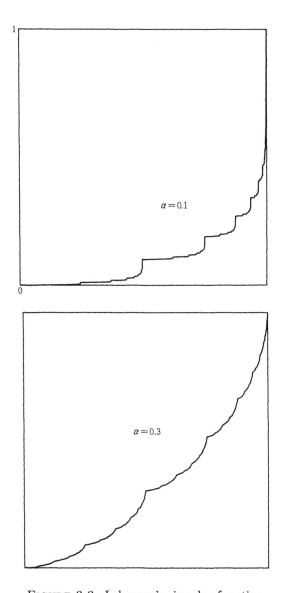

FIGURE 3.3. Lebesgue's singular function

where α is a number with $0 < \alpha < 1$. When $\alpha = 1/2$, we have a particularly simple solution, $M(x) = x$. Set $M_\alpha(x) = M(x)$ for $\alpha \neq 1/2$. Then $M_\alpha(x)$ is a unique continuous solution of (3.23) with the following properties (we prove this fact later in §3.4).

(i) $M_\alpha(x)$ is monotone increasing from 0 to 1.
(ii) The derivative of $M_\alpha(x)$ is almost everwhere zero.

We show the graph of this singular function of Lebesgue's in Figure 3.3 for the cases $\alpha = 0.1$ and $\alpha = 0.3$.

We can solve (3.23) rather easily. We begin with $x = 0$. The condition $\alpha \neq 1/2$ implies $M_\alpha(0) = \alpha M_\alpha(0)$, so that $M_\alpha(0) = 0$. For $x = 1$, $M_\alpha(1) = (1-\alpha)M_\alpha(1)+\alpha$ and so $M_\alpha(1) = 1$. For $x = 1/2$, $M_\alpha(1/2) = \alpha M_\alpha(1)$ and hence $M_\alpha(1/2) = \alpha$. Continuing this process well-defines all the $M_\alpha(i/2^k)$.

The function $M_\alpha(x)$ has the following statistical interpretation.

Imagine flipping a coin. With this coin one has the probability α of getting heads and the probability $1 - \alpha$ of getting tails. The game is unfair unless α is $1/2$.

We now determine a number ξ in the interval $[0, 1]$ by flipping the coin in infinitely many trials:

$$\xi = 0.\omega_0\omega_1\omega_2 \cdots = \frac{\omega_0}{2} + \frac{\omega_1}{2^2} + \cdots + \frac{\omega_n}{2^{n+1}} + \cdots,$$

where ω_n is 0 if it is heads and 1 if it is tails. We want to determine the probability distribution

$$f(t) = \mathrm{prob}\{\xi, \ \xi \geq t\}.$$

We set $\xi = (\omega_0 + \xi_1)/2$ to get

$$\xi_1 = \frac{\omega_1}{2} + \frac{\omega_2}{2^2} + \cdots.$$

On the other hand, ξ_1 has the same probability distribution $f(t)$, and moreover ω_0 and ξ_1 are two distinct independent variables.

The occurrence $\xi \leq t/2$ is the combination of the two occurrences $\omega_0 = 0$ and $\xi_1 < t$. Hence by the laws of probability we have

$$f\left(\frac{t}{2}\right) = \alpha f(t), \qquad 0 \leq t \leq 1.$$

Similarly, as the occurrence $\xi \leq (1 + t)/2$ is the combination of the occurrence $\omega_0 = 0$ or $\omega_0 = 1$ and the occurrence $\xi_1 \leq t$, we have

$$f\left(\frac{1+t}{2}\right) = \alpha + (1 - \alpha)f(t), \qquad 0 \leq t \leq 1.$$

The substitution $t \to 2x$ ($0 \leq x \leq 1/2$) and $t \to 2x - 1$ ($1/2 \leq x \leq 1$) in this formula results in (3.23). Indeed, if we set $f(t) = M_\alpha(t)$, we get

$$\begin{cases} M_\alpha(x) = \alpha M_\alpha(2x), & 0 \leq x \leq 1/2, \\ M_\alpha(x) = (1 - \alpha)M_\alpha(2x - 1) + \alpha, & \frac{1}{2} \leq x \leq 1, \end{cases}$$

which is the desired distribution function.

While we postpone the proof of the properties of Lebesgue's singular function, we use the following example to show that the definition of a curve given by (3.23) is indeed equivalent to the definition of a self-similar set we discussed in Chapter Two.

Let α be the complex number $\alpha = 1/2 + i\sqrt{3}/6$ and consider the following de Rham equation:

(3.24) $$\begin{cases} N_\alpha(x) = \alpha\overline{N_\alpha(2x)}, & 0 \leq x \leq \frac{1}{2}, \\ N_\alpha(x) = (1 - \alpha)\overline{N_\alpha(2x - 1)} + \alpha, & \frac{1}{2} \leq x \leq 1, \end{cases}$$

where $\overline{N_\alpha(2x)}$ is the complex conjugate of $N_\alpha(2x)$. This equation actually defines the Koch curve, shown in Figure 2.1 in Chapter Two. We see this by substituting consecutively, just as we did for (3.23), the values $x = 0, 1, 1/2, \ldots$, in (3.24) and looking at $N_\alpha(0), N_\alpha(1), N_\alpha(1/2), \ldots$, where the reader might feel uneasy about $x = 1/2$, but luckily both equations give the same value. The equations (3.24) correspond to the following two contractions of the complex plane:

$$F_0 : \ z_1 = \alpha \bar{z},$$
$$F_1 : \ z_1 = (1 - \alpha)\bar{z} + \alpha.$$

The fixed points of F_0 and F_1 are 0 and 1 respectively; however, the equations define a continuous curve since $F_0(1) = F_1(0) = \alpha$. As early as in 1957 de Rham proved this fact in a slightly more general setting: Let F_0 and F_1 be two contractions of the plane having p_0 and p_1 as their respective fixed points, and assume that $F_0(p_1) = F_0(p_0)$. Then the equations

(3.25)
$$\begin{cases} f\left(\frac{t}{2}\right) & = F_0(f(t)), \\ f\left(\frac{1+t}{2}\right) & = F_1(f(t)), \end{cases} \quad 0 \leq t \leq 1,$$

or equivalently

(3.25′)
$$\begin{cases} f(t) = F_0(f(2t)), & 0 \leq t \leq \frac{1}{2}, \\ f(t) = F_1(f(2t - 1)), & \frac{1}{2} \leq t \leq 1, \end{cases}$$

define a unique continuous curve connecting p_0 and p_1.

It is almost evident that the above equations give a parametrization of a self-similar set of Chapter Two.

3.4. The system of difference equations of Lebesgue's function

In §3.3 we derived Lebesgue's singular function M_α as the solution of (3.23). It is, just like the Takagi function, the solution to a boundary value problem with a system of infinitely many difference equations. In actual terms we have the following expression:

(3.26)
$$M_\alpha\left(\frac{2i + 1}{2^{k+1}}\right) = (1 - \alpha)M_\alpha\left(\frac{i}{2^k}\right) + \alpha M_\alpha\left(\frac{i + 1}{2^k}\right),$$
$$0 \leq i \leq 2^k - 1, \ k = 1, 2, \ldots, n, \ldots,$$

with the boundary values $M_\alpha(0) = 0$ and $M_\alpha(1) = 1$.

REMARK. The only difference between $M_\alpha(x)$ and $T(x)$ is that $T(x)$ satisfies the homogeneous boundary condition $T(0) = T(1) = 0$ of a non-homogeneous system while $M_\alpha(x)$ satisfies the non-homogeneous boundary condition of a homogeneous system.

The derivation of (3.26) follows easily from (3.23) for the following reason. The respective numerators i and $i + 1$ of $i/2^k$ and $(i + 1)/2^k$ both lie in $[0, 2^{k-1}]$, or in $[2^{k-1}, 2^k - 1]$. Hence $(2i + 1)/2^{k+1}$ belongs to $[0, 1/2]$ or to $[1/2, 1]$ accordingly.

In the first case, using (3.23) we get

$$M_\alpha\left(\frac{i}{2^k}\right) = \alpha M_\alpha\left(\frac{2i}{2^k}\right) = \alpha M_\alpha\left(\frac{i}{2^{k-1}}\right),$$

$$M_\alpha\left(\frac{i+1}{2^k}\right) = \alpha M_\alpha\left(\frac{2i+2}{2^k}\right) = \alpha M_\alpha\left(\frac{i+1}{2^{k-1}}\right),$$

$$M_\alpha\left(\frac{2i+1}{2^{k+1}}\right) = \alpha M_\alpha\left(\frac{2i+1}{2^k}\right),$$

and in the second case we have

$$M_\alpha\left(\frac{i}{2^k}\right) = (1-\alpha)M_\alpha\left(\frac{2i}{2^k}-1\right) + \alpha = (1-\alpha)M_\alpha\left(\frac{i-2^{k-1}}{2^{k-1}}\right) + \alpha,$$

$$M_\alpha\left(\frac{i+1}{2^k}\right) = (1-\alpha)M_\alpha\left(\frac{2(i+1)}{2^k}-1\right) + \alpha = (1-\alpha)M_\alpha\left(\frac{i+1-2^{k-1}}{2^{k-1}}\right) + \alpha,$$

$$M_\alpha\left(\frac{2i+1}{2^{k+1}}\right) = (1-\alpha)M_\alpha\left(\frac{2i+1}{2^k}-1\right) + \alpha = (1-\alpha)M_\alpha\left(\frac{i+\frac{1}{2}-2^{k-1}}{2^{k-1}}\right) + \alpha.$$

Notice on the other hand that

$$\frac{i-2^{k-i}+i+1-2^k}{2} = \frac{2i+1-2\cdot 2^{k-1}}{2} = i + \frac{1}{2} - 2^{k-1}.$$

Now when $k=1$, we have $0 \le i \le 1$, and so $i = 0$ or 1. We then get

$$\alpha^2 = M_\alpha\left(\frac{1}{4}\right) = \alpha M_\alpha\left(\frac{1}{2}\right) + (1-\alpha)M_\alpha\left(\frac{0}{2}\right) = \alpha^2,$$

$$2\alpha - \alpha^2 = M_\alpha\left(\frac{3}{4}\right) = (1-\alpha)M_\alpha\left(\frac{1}{2}\right) + \alpha M_\alpha(1) = 2\alpha - \alpha^2.$$

We complete the proof by induction, starting at $k = 1$.

REMARK. We may rewrite the system of difference equations (3.26) as follows:

(3.27)
$$\begin{aligned} M_\alpha\left(\frac{2i+1}{2^{k+1}}\right) &= \frac{1}{2}\left\{M_\alpha\left(\frac{i}{2^k}\right) + M_\alpha\left(\frac{i+1}{2^k}\right)\right\} \\ &\quad + \left(\alpha - \frac{1}{2}\right)\left\{M_\alpha\left(\frac{i+1}{2^k}\right) - M_\alpha\left(\frac{i}{2^k}\right)\right\}. \end{aligned}$$

We see by dividing both sides by $2^{2(k+1)}$ that this equation is a generalization of the discrete singular perturbation problem.

We utilize the above calculation to obtain, in the reverse direction from the case of the Takagi function, Schauder's expansion of M_α. Using (3.16) we find the Schauder coefficient $a_{k,i}$ of M_α to be

(3.28)
$$a_{k,i} = \left(\alpha - \frac{1}{2}\right)\left\{M_\alpha\left(\frac{i+1}{2^k}\right) - M_\alpha\left(\frac{i}{2^k}\right)\right\};$$

hence, in order to find the difference quotient,

$$\left\{M_\alpha\left(\frac{i+1}{2^k}\right) - M_\alpha\left(\frac{i}{2^k}\right)\right\}\bigg/ \frac{1}{2^k},$$

it is enough to calculate $a_{k,i}$.

We treat the two cases $i = 2l$ and $i = 2l + 1$ separately, but in both cases we use (3.23) and (3.28) repeatedly. We have, for $i = 2l$,

$$a_{k,2l} = \alpha a_{k-1,l},$$

and for $i = 2l + 1$,

$$a_{k,2l+1} = (1 - \alpha)a_{k-1,l}.$$

Furthermore, as $a_{0,0} = \alpha - 1/2$, we have

THEOREM 3.2. *The coefficient $a_{k,i}$ of Schauder's expansion of $M_\alpha(x)$ has the expression*

$$a_{k,i} = \left(\alpha - \frac{1}{2}\right)\alpha^p(1 - \alpha)^q.$$

Here $p + q = k$ with p the number of zeros and q the number of ones showing up among $\omega_1 \cdots \omega_k$ in the binary representation

$$i = \sum_{j=1}^{k}\omega_j \cdot 2^{j-1}, \qquad 0 \le i \le 2^k - 1.$$

The function $M_\alpha(x)$, therefore, has the following Schauder expansion:

$$M_\alpha(x) = x + \sum_{k=0}^{\infty}\sum_{i=0}^{2^k-1}\left(\alpha - \frac{1}{2}\right)\alpha^p(1 - \alpha)^q F_{\frac{i}{2^k}, \frac{i+1}{2^k}}(x).$$

Notice in particular that $M_\alpha(x)$ is a regular function of α.

We now apply the above formula to verify the properties of $M_\alpha(x)$ as promised in §3.3. We use (3.28) to derive the inequality

(3.29)
$$M_\alpha\left(\frac{i+1}{2^k}\right) - M_\alpha\left(\frac{i}{2^k}\right) = \alpha^p(1 - \alpha)^q > 0$$

for every i and every k, which verifies the property (i) of $M_\alpha(x)$.

The function $M_\alpha(x)$ is monotone increasing in the strict sense. It is thus a function of bounded variation, and so its derivative exists almost everywhere. Suppose that $M_\alpha(x)$ is differentiable at a point x_0 in the interval $[0, 1]$. Then for each k there exists some j_k such that

$$\frac{j_k}{2^k} \le x_0 \le \frac{j_k + 1}{2^k},$$

$$\frac{M_\alpha\left(\frac{j_k+1}{2^k}\right) - M_\alpha\left(\frac{j_k}{2^k}\right)}{2^{-k}} = \alpha^p(1 - \alpha)^q \cdot 2^k \equiv D_k \cdot 2^k,$$

where we assume that $\lim_{k\to\infty} D_k \cdot 2^k$ has a finite value different from zero. On the other hand, the following must hold:

$$\frac{2^{k+1}D_{k+1}}{2^k D_k} = 2\frac{\alpha^{p'}(1 - \alpha)^{q'}}{\alpha^p(1 - \alpha)^q} = 2\alpha \quad \text{or} \quad 2(1 - \alpha),$$

$$\lim_{k\to\infty}\frac{2^{k+1}D_{k+1}}{2^k D_k} = 1,$$

which amounts to

$$\lim_{k\to\infty}\frac{D_{k+1}}{D_k} = \frac{1}{2}.$$

Hence, we must have

$$\frac{D_{k+1}}{D_k} = \alpha \quad \text{or} \quad 1 - \alpha,$$

which, together with our choice of $\alpha \neq 1/2$, implies that

$$\lim_{k \to \infty} 2^k D_k = 0.$$

But this is a contradiction, as we assumed that this limit was not zero. Thus we conclude that the derivative of $M_\alpha(x)$ is zero wherever it exists.

3.5. The relation between $T(x)$ and $M_\alpha(x)$ and its generalization

We would like to show that the respective functions of Takagi and Lebesgue we have been studying are related in an amazingly simple way. Let us first consider the system of difference equations for $M_\alpha(x)$ and then differentiate each equation with respect to α. So we start out with

$$M_\alpha \left(\frac{2i+1}{2^{k+1}} \right) = (1-\alpha)M_\alpha \left(\frac{i}{2^k} \right) + \alpha M_\alpha \left(\frac{i+1}{2^k} \right),$$

$$M_{\alpha+h} \left(\frac{2i+1}{2^{k+1}} \right) = (1-\alpha-h)M_{\alpha+h} \left(\frac{i}{2^k} \right) + (\alpha+h)M_{\alpha+h} \left(\frac{i+1}{2^k} \right),$$

take the difference of these two equations, and divide it by h:

$$(1-\alpha) \left(\frac{M_{\alpha+h} - M_\alpha}{h} \right) \left(\frac{i}{2^k} \right) + \alpha \left(\frac{M_{\alpha+h} - M_\alpha}{h} \right) \left(\frac{i+1}{2^k} \right)$$
$$+ M_{\alpha+h} \left(\frac{i+1}{2^k} \right) - M_\alpha \left(\frac{i}{2^k} \right) = \left(\frac{M_{\alpha+h} - M_\alpha}{h} \right) \left(\frac{2i+1}{2^{k+1}} \right).$$

Now assume $\alpha = 1/2$ and let $h \to 0$; then, incorporating (3.29) as well, we can evaluate

$$\lim_{h \to 0} \frac{M_{\alpha+h} - M_\alpha}{h} = \left. \frac{\partial M_\alpha}{\partial \alpha} \right|_{\alpha = \frac{1}{2}}$$

at $x = \dfrac{2i+1}{2^{k+1}}$ as follows:

$$\frac{1}{2} \left\{ \left(\left. \frac{\partial M_\alpha}{\partial \alpha} \right|_{\alpha = \frac{1}{2}} \right) \left(\frac{i}{2^k} \right) + \left(\left. \frac{\partial M_\alpha}{\partial \alpha} \right|_{\alpha = \frac{1}{2}} \right) \left(\frac{i+1}{2^k} \right) \right\} + \frac{1}{2^k}$$
$$= \left(\left. \frac{\partial M_\alpha}{\partial \alpha} \right|_{\alpha = \frac{1}{2}} \right) \left(\frac{2i+1}{2^{k+1}} \right),$$

which in turn satisfies the system of difference equations (3.19) having the Takagi function as its solution:

$$\frac{1}{2} \left\{ T \left(\frac{i}{2^k} \right) + T \left(\frac{i+1}{2^k} \right) \right\} + \frac{1}{2^{k+1}} = T \left(\frac{2i+1}{2^{k+1}} \right).$$

Since the solution of each of these two systems is unique, we have arrived a very simple relation:

$$(3.30) \qquad \qquad \left. \frac{\partial M_\alpha}{\partial \alpha} \right|_{\alpha = \frac{1}{2}} = 2T(x).$$

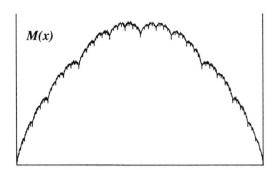

FIGURE 3.4. An example of a complicated Takagi function

The above result and its generalization are included in a paper published by Hata and Yamaguti in 1984[2]. The following theorems summarize the ideas in our paper. We begin with a theorem which extends the Takagi function. This theorem also generalizes the Dirichlet problem for the Poisson equation.

THEOREM 3.3. *The system of equations*

(3.31)
$$f\left(\frac{2i+1}{2^{k+i}}\right) - \frac{1}{2}\left\{f\left(\frac{i}{2^k}\right) + f\left(\frac{i+1}{2^k}\right)\right\} = c_k,$$
$$0 \le i \le 2^k - 1, \ k = 0, 1, 2, \ldots,$$
$$f(0) = f(1) = 0,$$

has a unique continuous solution $f(x)$ if and only if

(3.32)
$$\sum_{n=0}^{\infty} |c_n| < +\infty.$$

We named the set of the solutions appearing in Theorem 3.3 the Takagi class, among which the reader will find, for instance, by taking $c_n = 1/(1 + n^2)$, the complicated function whose graph appears in Figure 3.4. We have not succeeded in computing the Hausdorff dimensions of the graphs of functions in the Takagi class other than the Takagi function itself.

The proof of the sufficiency of Theorem 3.3 is almost trivial, and was given by Faber in 1906, while the necessity part is difficult, demanding the use of the fact that the orbits of the discrete dynamical system

$$x_{n+1} = \varphi(x_n)$$

are chaotic. We encourage the reader to read our original paper.

We have the following theorem for generalizing the Lebesgue function.

---

[2]Hata, M., and Yamaguti, M., Takagi function and its generalization, Japan J. Appl. Math., **1**(1984), 186–199.

THEOREM 3.4. *The system of difference equations*

$$f\left(\frac{2i+1}{2^{k+1}}\right) = (1-\alpha)f\left(\frac{i}{2^k}\right) + \alpha f\left(\frac{i+1}{2^k}\right) + c_k,$$

$$0 < \alpha < 1, \ 0 \le i \le 2^k - 1, \ k = 0, 1, 2, \ldots,$$

$$f(0) = 0, \quad f(1) = 1,$$

has a unique solution if $\sum_{n=0}^{\infty} |c_n| < +\infty.$

In 1991 Sekiguchi and Shiota[3] showed that $\sum_{n=0}^{\infty} |c_n| < +\infty$ is the necessary condition for the above theorem. They generalized, moreover, the relation (3.30) and computed the l-th partial derivative of $M_\alpha(x)$ with respect to α (cf. the references). We also generalized Schauder's expansion discussed in §3.2 and proposed a new expansion which has a basis such that the function

$$f\left(\frac{2i+1}{2^{k+1}}\right) - (1-\alpha)f\left(\frac{i}{2^k}\right) - \alpha f\left(\frac{i+1}{2^k}\right)$$

of Theorem 3.4 corresponds to the coefficient of $i/2^k$. For instance, the second term of this new expansion is $M_\alpha(x)$ whereas that of Schauder's is x. We also replaced the function (3. 13) by a quadratic function and investigated how this affected the coefficients of Schauder's expansion[4]. Then Kigami came up with the following family of basis functions:

$$(3.33) \qquad \xi_t(x) = \sum_{n=1}^{\infty} t^{n-1} \varphi^n(x), \qquad 0 \le t < 1, \quad 0 \le x \le 1.$$

By (3.11) the function ξ_t is $4x(1-x)$ for $t = 1/4$, and $2T(x)$ for $t = 1/2$. We define our basis elements to be

$$\xi_{k,i}^t(x) = \begin{cases} \xi_t\left(2^k\left(x - \frac{i}{2^k}\right)\right), & \frac{i}{2^k} \le x \le \frac{i+1}{2^k}, \\ 0, & \frac{i}{2^k} > x \text{ or } x > \frac{i+1}{2^k}. \end{cases}$$

The expansion of a continuous function $f(x)$ in the above basis has the form

$$f(x) = c_0 + c_1 x + \sum_{k=0}^{\infty} \sum_{i=0}^{2^k-1} c_{k,i} \xi_{k,i}^t(x),$$

where the coefficients $c_{k,i}$ have a very simple relation with the Schauder coefficients:

$$(3.34) \qquad c_{k,i} = \begin{cases} a_{k,i} - t a_{k-1,\frac{i}{2}}, & i \text{ even}, \\ a_{k,i} - t a_{k-1,\frac{i-1}{2}}, & i \text{ odd}. \end{cases}$$

---

[3]Sekiguchi, T., and Shiota, Y., A generalization of Hara–Yamaguti's results on the Takagi function, Japan J. Indust. Appl. Math., 8(1991), 203–219.

[4]Yamaguti, M., Schauder expansion by some quadratic base function, J. Fac. Sci., Univ. Tokyo, Sect. IA Math., 36(1989), 187–191.

3.6. Wavelet expansions

What characterizes a Schauder expansion is this: we adjust a certain appropriete function, say $F_{\alpha,\beta}(x)$, to define a family of new functions in such a way that each subinterval $[i/2^k, (i+1)/2^k]$ becomes a support of one of them, and we use all of them in the expansion. Some call this method the degree analysis of a multisolution image. It differs from the Fourier expansion in that $i/2^k$ *corresponding to the period n contains some information pertaining to position in x-space.* Furthermore, as in the case of Fourier expansions, if $F_{\alpha,\beta}$ is orthogonal to every other $F_{\alpha',\beta'}$, then we can find the coefficients by integration.

Another difference is that we can read a sudden change in the behavior of the function $f(x)$, say the discontinuity of its derivative for example, from the size of the coefficients of the expansion. This will be a big advantage over the Fourier expansion, which cannot locate the points where the original function behaves singularly.

We do have a basis which satisfies the above orthogonality condition. We consider the derivative $\varphi'(x)$ of $\varphi(x)$ to define a basis called the Haar basis:

$$H_{k,i}(x) = 2^{k/2} H(2^k x - i), \qquad 0 \le i \le 2^k - 1, \ k = 0, 1, 2, \ldots,$$

where $H(x)$ is defined by

$$H(x) = \frac{1}{2}\varphi'(x) = \begin{cases} 1, & 0 \le x < \frac{1}{2}, \\ -1, & \frac{1}{2} \le x \le 1. \end{cases}$$

Note that $\int H_{k,i}(x)\,dx = 0$; the average is zero. We call $H(x)$ Haar's wavelet.

Furthermore, when $(k,i) \ne (k',i')$ we have the orthogonality of $H_{k,i}(x)$:

$$\int H_{k,i}(x) H_{k',i'}(x)\,dx = 0.$$

Hence we can define a complete orthonormal basis which generates the space of square-integrable functions. Another very good thing about this expansion is that, just like Fourier coefficients, the coefficients of the expansion of $f(x)$,

$$f(x) = \sum_{k=0}^{\infty} \sum_{i=0}^{2^k-1} a_{k,i} H_{k,i}(x),$$

are computable. We have in fact

$$a_{k,i} = \int_{-\infty}^{+\infty} f(x) H_{k,i}(x)\,dx.$$

In addition, the coefficient $a_{k,i}$ becomes large as k becomes large at the points where $f(x)$ is discontinuous with an order of magnitude much larger than what it is where $f(x)$ is smoothly continuous. So the coefficicients of the expansion help us locate the points of discontinuity.

If we wish to proceed further and specify where even the higher order derivatives of $f(x)$ are discontinuous, then we would have to insist that the above $H(x)$ be smooth and that all moments be zero. On top of that, we must require their orthogonality as well. In order to deal with all these requirements, we have currently two types of wavelets under consideration.

(a) Meyer's analyzing wavelets. First we construct $\phi(x)$. Denote by $\tilde{\phi}(\xi)$ the Fourier transform of $\phi(x)$, defined as follows:

$$\tilde{\phi}(\xi) = \int_{-\infty}^{+\infty} \phi(x)e^{-ix\xi}\,dx,$$

subject to the conditions:

(i) $\tilde{\phi}(\xi) \geq 0$, $-\infty < \xi < +\infty$, and $\tilde{\phi}(\xi)$ is monotone increasing for $\xi \geq 0$.

(ii) $\tilde{\phi}(\xi) = \tilde{\phi}(-\xi)$.

(iii) $\tilde{\phi}(\xi) = 1$, $|\xi| \leq \frac{2}{3}\pi$; $\tilde{\phi}(\xi) = 0$, $|\xi| \geq \frac{4}{3}\pi$.

(iv) $(\tilde{\phi}(\xi))^2 + (\tilde{\phi}(\xi - 2\pi))^2 = 1$, $\frac{2}{3}\pi \leq \xi \leq \frac{4}{3}\pi$.

We now construct a new function using this $\phi(x)$. First define $\tilde{\psi}(x)$ by

(3.35) $$\tilde{\psi}(x) = e^{-i\xi/2}\sqrt{(\tilde{\phi}(\xi/2))^2 - (\tilde{\phi}(\xi))^2},$$

and then inverse–Fourier–transform it to define $\psi(\xi)$:

(3.36) $$\psi(\xi) = \frac{1}{2\pi}\int_{-\infty}^{+\infty} e^{i\xi x}\tilde{\psi}(\xi)\,d\xi,$$

which we call Meyer's *analyzing wavelet*. For example, we might construct such a function $\tilde{\phi}(\xi)$ as follows. We first define a function $\gamma(\xi)$ by

$$\gamma(\xi) = \frac{\int_0^\xi \{t(1-t)\}^n\,dt}{\int_0^1 (1-t)^n\,dt}.$$

Then $\gamma(\xi) \geq 0$ for $0 \leq \xi \leq 1$, $\gamma(0) = 0$ and $\gamma(1) = 1$. Define $\tilde{\phi}(\xi)$ by

$$\tilde{\phi}(\xi) = \begin{cases} 0, & \xi \leq -\frac{4}{3}\pi, \\ \sqrt{\gamma\left(\frac{3}{2\pi}\xi + 2\right)}, & -\frac{4}{3}\pi \leq \xi \leq -\frac{2}{3}\pi, \\ 1, & -\frac{2}{3}\pi \leq \xi \leq \frac{2}{3}\pi, \\ \sqrt{\gamma\left(2 - \frac{3}{2\pi}\xi\right)}, & \frac{2}{3}\pi \leq \xi \leq \frac{4}{3}\pi, \\ 0, & \xi \geq \frac{4}{3}\pi. \end{cases}$$

Then $\tilde{\phi}(\xi)$ satisfies (i), (ii), (iii), and (iv); therefore, plugging this into (3.36), we get a desired wavelet $\psi(\xi)$, which we show in Figure 3.5. We can of course construct wavelets other than $\psi(\xi)$.

Note that the property of $\phi(x)$ in the following proposition shows that $\psi(\xi)$ is indeed what we wanted.

PROPOSITION. *The function $\phi(x)$ is infinitely differentiable, and belongs to the set $L^1(\mathbf{R}) \cap L^2(\mathbf{R}) \cap L^\infty(\mathbf{R})$. Further, $\phi^{(n)}(x) \in L^1(\mathbf{R}) \cap L^2(\mathbf{R})$ for each n, and $\phi(x)$ defines a complete orthonormal basis of $L^2(\mathbf{R})$.*

We only show the orthonormality of the functions $\phi(x - k)$; that is, in the subspace spanned by $\phi(x - k)$, $k = 0, \pm 1, \pm 2, \ldots, \pm n, \ldots$, they are orthonormal. Note that the inner product $\langle\ \rangle$ for $L^2(\mathbf{R})$ is the usual

$$\langle \phi(x - k), \phi(x - l) \rangle = \int_{-\infty}^{+\infty} \phi(x - k)\overline{\phi}(x - l)\,dx,$$

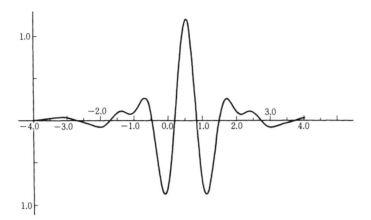

FIGURE 3.5. Meyer's wavelet

which becomes, by Parseval's equality,

$$\langle \phi(x-k), \phi(x-l) \rangle = \frac{1}{2\pi} \int_{-\infty}^{+\infty} |\tilde{\phi}(\xi)|^2 e^{i(k-l)\xi} \, d\xi.$$

Noticing that $e^{i(k-l)\xi}$ is a function of period 2π, we get

$$\langle \phi(x-k), \phi(x-l) \rangle = \frac{1}{2\pi} \int_{0}^{2\pi} \sum_{m=-\infty}^{+\infty} |\tilde{\phi}(\xi + 2m\pi)|^2 e^{-i(k-l)\xi} \, d\xi$$

$$= \delta_{k,l} \text{ (by property (iv) of } \tilde{\phi}(\xi)).$$

Let V_0 be the closure of the subspace of L^2 spanned by $\phi(x-k)$, $k = 0, 1, 2, \ldots$; so V_0 is a closed subspace of L^2. We define inductively a sequence of closed subspaces V_j of L^2, $j = 0, 1, 2, \ldots$, as follows:

$$f(x) \in V_j \iff f(2x) \in V_{j+1},$$

where V_0 is as above.

We can show that the family of functions $\{2^{j/2}\phi(2^j x - k)\}$ forms a complete orthonormal basis and that $V_j \subset V_{j+1}$, so that $\bigcup_{-\infty}^{+\infty} V_j = L^2(\mathbf{R})$ and $\bigcap_{-\infty}^{+\infty} V_j = \{0\}$. Furthermore, from

$$\tilde{\psi}(\xi) = e^{-i\xi/2} \sqrt{(\tilde{\phi}(\xi/2))^2 - (\tilde{\phi}(\xi))^2}$$

it follows that, on setting $\psi_{j,k} = 2^{j/2}\psi(2^j x - k)$, $\psi(x)$ too generates a complete orthonormal basis of $L^2(\mathbf{R})$.

The wavelet $\psi(x)$ enjoys the following three properties for $\tilde{\phi}(\xi) \in C^k(\xi)$:

1°. $\psi(x) = O(|x|^{-k})$, $|x| \to +\infty$.

2°. $\psi(x) \in C^\infty(\mathbf{R})$, and $\psi^{(n)}(x) = O(|x|^{-k})$, $|x| \to +\infty$.

3°. $\int_{-\infty}^{+\infty} x^l \psi(x) \, dx = 0$, $l = 0, 1, 2, \ldots, k$.

(b) Daubechies's wavelets. Unlike Haar's wavelet, the supports of the analyzing wavelets discussed above are not bounded. For this reason it does not commute with differential operators, and in this respect it is inferior to the Fourier transform. Daubechies, in an effort to improve this situation, succeeded in smoothing out the Haar basis in 1988. Even though we cannot assert that $\psi(x) \in C^\infty(\mathbf{R})$, we can choose $\psi(x)$ equipped with the regularity of a desired rank.

Suppose for any natural number N we can construct a sequence

$$h(0), h(1), \ldots, h(2N - 1),$$

with the following properties:

(i) $\sum_k h(k)h(k + 2m) = \delta_{0,m}$ for every integer m; here we set $h(k + 2m) = 0$ if $k + 2m \neq 0, 1, \ldots,$ or $2N - 1$.

(ii) $\sum_k h(k) = \sqrt{2}$, and if we set $g(k) = (-1)^k h(-k + 1)$, then

$$\sum_k g(k)k^m = 0, \qquad 0 \leq m < N.$$

We then build $\phi(x)$ as well as $\psi(x)$ on the finite sequences $\{h(i)\}$ and $\{g(i)\}$:

$$m_0(\xi) = \frac{1}{\sqrt{2}} \sum_k h(k)e^{ik\xi},$$

$$\tilde{\phi}(\xi) = m_0(\xi)\tilde{\phi}\left(\frac{\xi}{2}\right) = \prod_{k \geq 0} m_0(2^{-k}\xi),$$

$$\psi(x) = \sqrt{2} \sum_k g(k)\phi(2x - k).$$

When $N = 1$ we have a Haar's wavelet with

$$h(0) = \frac{1}{\sqrt{2}}, \quad h(1) = \frac{1}{\sqrt{2}}, \quad g(0) = \frac{1}{\sqrt{2}} \quad g(1) = -\frac{1}{\sqrt{2}},$$

and

$$m_0(\xi) = \left(\frac{1}{2} + \frac{1}{2}e^{i\xi/2}\right) = \frac{1}{\sqrt{2}}(h(0) + h(1)e^{i\xi}).$$

In the Haar environment, $\phi(x)$ satisfies $\phi(x) = \phi(2x) + \phi(2x - 1)$, and therefore the equality $\tilde{\phi}(\xi) = m_0(\xi)\tilde{\phi}(\xi/2)$ holds. Thus we have $\psi(x) = \phi(2x) - \phi(2x - 1)$.

Daubechies explicitly calculated the properties (i) and (ii) up to $N = 10$ and constructed the corresponding wavelets. If we define, in the same way we did Meyer's analyzing wavelet, $\phi_{k,i}$, $\psi_{k,i}$, and then V_k, $W_k = L^2(\mathbf{R}) - V^k$, we obtain a complete orthonormal basis for $L^2(\mathbf{R})$. Furthermore:

$$\text{Support of } \phi_{n,k} = [2^{-n}, 2^{-n}(k + 2n - 1)],$$

$$\text{Support of } \psi_{n,k} = [2^{-k}(k + 1 - n), \tau^n(k + n)],$$

$$\int \psi_{j,k}(x)x^m \, dx = 0, \qquad 0 \leq m < N - 1.$$

We mention that every $\phi_{j,k}$ and every $\psi_{j,k}$ is Hölder continuous of degree $\lambda(N)$; for instance, $\lambda(2) = 0.55$, $\lambda(3) \approx 1.087$, and $\lambda(4) = 1.6$. We show two of Daubechies's wavelets in Figure 3.6.

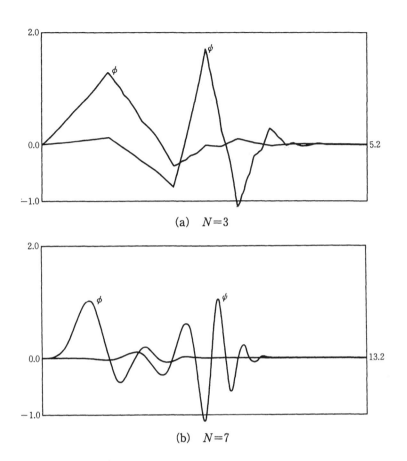

(a) $N=3$

(b) $N=7$

FIGURE 3.6. Daubechies's wavelets

Exercises

3.1. Show that $T(x)$ of §1.4 and $T(x)$ of §3.1 represent the same Takagi function.

3.2. Show that the de Rham's system of equations (3.25) has a unique continuous solution $f(x)$.

3.3. Show that Schauder's expansion (3.15) converges uniformly in the interval $[0, 1]$.

3.4. Prove the sufficiency part (3.32) of Theorem 3.3.

3.5. Using a wavelet of J. Morlet

$$\psi(t) = \begin{cases} 0, & t \le 0, \\ t, & 0 \le t \le 1/3, \\ 1 - 2t, & 1/3 \le t \le 2/3, \\ t - 1, & 2/3 \le t \le 1, \\ 0, & t \ge 1, \end{cases}$$

define a base function ψ_I by

$$\psi_I = \frac{3\sqrt{3}}{\sqrt{h}} \psi \left(\frac{t-a}{h} \right), \qquad I = [a, b], \quad b - a = h,$$

and expand the function $f(t) = \alpha_1 \psi_{I_1} + \alpha_2 \psi_{I_2} + \cdots$. Assume that one can orthogonalize ψ_{I_i}, $i = 1, 2, \ldots$, for all i. Then for $\xi \in I_i$, show that the wavelet coefficient

$$\alpha(I) = \int_{-\infty}^{+\infty} f(t) \psi_I(t) \, dt$$

at ξ is $O(\sqrt{h})$ if ξ is a discontinuity point of $f(t)$ of the first kind, and it is equal to $O(h\sqrt{h})$ if ξ is a point of continuity.

CHAPTER 4

In Quest of Fractal Analysis

In Chapters One and Two we discussed the geometrical properties (Haudorff dimension, connectivity, self-similarity, *etc.*) of fractals. The concept of fractals came to represent objects in nature in place of traditional smooth sets such as circles, spheres, straight lines, ellipses, ... Hence, in order to analyze natural phenomena, we must study physical phenomena occurring in the objects represented by fractals. For example, one uses differential equations on smooth sets to describe the motion of waves and dispersion. How then do we translate these differential equations onto fractals? This is a problem never dealt with in the standard analysis. In this chapter, as a first step toward analysis on fractals, we construct an operator comparable to the Laplacian on a self-similar set called the Sierpiński gasket, where we discuss Green's formula, Poisson's equation, and the properties of harmonic functions. In §4.1, we define the Sierpiński gasket and construct harmonic functions from difference equations which correspond to Laplacians on it. In §4.2, we build a physical model for the wave equation on the Sierpiński gasket, and by the physical observation of this model we clarify how we should define the Laplacian on it. In §§4.3 and 4.4, we define the Laplacian and deal with the Gauss–Green type formula, Poisson's equation, the Dirichlet problem, *etc.*

4.1. The Sierpiński gasket

At the beginning of this century Sierpiński discovered a set which we call today the *Sierpiński gasket*.

DEFINITION 4.1. Let $\{p_1, p_2, p_3\}$ be the vertices of an equilateral triangle of unit length in the plane. For $i = 1, 2, 3$ define a contraction F_i of the plane by

$$F_i(x) = \frac{1}{2}(x - p_i) + p_i,$$

which has the fixed point p_i. Then the set of contractions $\{F_1, F_2, F_3\}$ defines a self-similar set K (cf. Definition 2.1 and Theorem 2.1). We call K the Sierpiński gasket; that is, K is a unique compact set satisfying

$$K = F_1(K) \cup F_2(K) \cup F_3(K).$$

We mention that the Hausdorff dimension of the Sierpiński gasket is $\log 3 / \log 2$ (Exercise 4.1).

In order to develop analysis on the Sierpiński gasket we define a sequence (V_m, B_m) of graphs approximating it, where V_m and B_m denote the set of vertices and the set of edges respectively, as follows (see Figure 4.1):

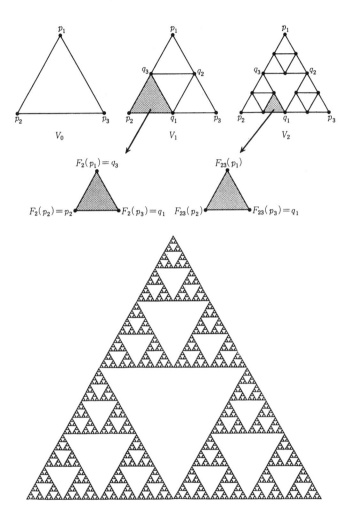

FIGURE 4.1. The Sierpiński gasket and its approximating sequence (V_m, B_m)

DEFINITION 4.2. For $V_0 = \{p_1, p_2, p_3\}$, $B_0 = \{(p_1, p_2), (p_1, p_3), (p_2, p_3)\}$, and each integer $m \geq 1$, we set

$$V_m = \bigcup_{1 \leq i_1, i_2, \dots, i_m \leq 3} F_{i_1} \circ F_{i_2} \circ \cdots \circ F_{i_m}(V_0),$$

$$B_m = \Big\{ \big(F_{i_1} \circ F_{i_2} \circ \cdots \circ F_{i_m}(p_k), F_{i_1} \circ F_{i_2} \circ \cdots \circ F_{i_m}(p_l)\big) \Big| \, 1 \leq i_1, \dots, i_m \leq 3,$$
$$1 \leq k < l \leq 3 \Big\}.$$

Set $V_* = \bigcup_{m \geq 0} V_m$. The closure $\overline{V}_*$ is the Sierpiński gasket. For future convenience we further adopt the following notation.

1. $\{1, 2, 3\}^m = \{w_1 w_2 \dots w_m | \, 1 \leq w_1, w_2, \dots, w_m \leq 3\}$.

2. For $w = w_1 w_2 \cdots w_m \in \{1, 2, 3\}^m$,

$$F_w = F_{w_1} \circ F_{w_2} \circ \cdots \circ F_{w_m}, \quad K_w = F_{w_1} \circ F_{w_2} \circ \cdots \circ F_{w_m}(K),$$

$$p_i(w) = F_w(p_i), \quad q_i(w) = F_w(q_i), \quad i = 1, 2, 3.$$

With this notation we may write

$$V_m = \bigcup_{w \in \{1,2,3\}^m} F_w(V_0), \qquad F_w(V_0) = \{p_1(w), p_2(w), p_3(w)\}.$$

REMARK. Let ω be a one-sided infinite sequence of the numbers 1, 2 and 3: $\omega = \omega_1 \omega_2 \omega_3 \cdots$, $\omega_i = 1, 2$ or 3. Then we have $K_{\omega_1 \omega_2 \cdots \omega_m} \supset K_{\omega_1 \omega_2 \cdots \omega_m \omega_{m+1}}$ for every $m \geq 1$. Furthermore, as the diameter of the set $K_{\omega_1 \cdots \omega_m \omega_{m+1}}$ is $(1/2)^m$, the set $\bigcap_{m \geq 1} K_{\omega_1 \omega_2 \cdots \omega_m}$ is a singleton. This correspondence defines a map π from the set of the one-sided infinite sequences $\Sigma = \{\omega \mid \omega = \omega_1 \omega_2 \cdots, \ \omega = 1, 2, 3\}$ to K:

$$\{\pi(\omega)\} = \bigcap_{m \geq 1} K_{\omega_1 \omega_2 \cdots \omega_m}.$$

The map $\pi : \Sigma \to K$ is surjective. Also, π is continuous with respect to the product topology of Σ as the infinite product of $\{1, 2, 3\}$. We evaluate π for some specific points in Σ:

$$\pi(\dot{1}) = p_1, \quad \pi(\dot{2}) = p_2, \quad \pi(\dot{3}) = p_3, \quad \pi(1\dot{2}) = \pi(2\dot{1}) = q_3,$$

$$\pi(1\dot{3}) = \pi(3\dot{1}) = q_2, \quad \pi(2\dot{3}) = \pi(3\dot{2}) = q_1,$$

where $\dot{1} = 1111 \cdots$, $\dot{2} = 2222 \cdots$, $\dot{3} = 3333 \cdots$.

We can construct in a similar way a map π for more general self-similar sets, which is quite important in studying the topology of self-similar sets[1].

We now define the discrete Laplacian on the graph (V_m, B_m).

DEFINITION 4.3. We set $l(V_m) = \{f \mid f : V_m \to \mathbf{R}\}$, and define a map $H_m : l(V_m) \to l(V_m)$ by

$$(H_m f)(p) = \sum_{q \in V_{m,p}} (f(q) - f(p)),$$

where $f \in l(V_m)$, $p \in V_m$, and $V_{m,p}$ denotes the set

$$V_{m,p} = \{q \mid q \text{ is connected to } p \text{ by an edge in } (V_m, B_m)\}$$
$$= \{q \mid (p, q) \in B_m \text{ or } (q, p) \in B_m\}.$$

In terms of matrices we have the following:

$$H_0 = \begin{pmatrix} -2 & 1 & 1 \\ 1 & -2 & 1 \\ 1 & 1 & -2 \end{pmatrix}, \quad \text{and} \quad \begin{pmatrix} (H_0 f)(p_1) \\ (H_0 f)(p_2) \\ (h_0 f)(p_3) \end{pmatrix} = H_0 \begin{pmatrix} f(p_1) \\ f(p_2) \\ f(p_3) \end{pmatrix}.$$

If we set

$$T = \begin{pmatrix} -2 & 0 & 0 \\ 0 & -2 & 0 \\ 0 & 0 & -2 \end{pmatrix}, \quad J = \begin{pmatrix} 0 & 1 & 1 \\ 1 & 0 & 1 \\ 1 & 1 & 0 \end{pmatrix}, \quad X = \begin{pmatrix} -4 & 1 & 1 \\ 1 & -4 & 1 \\ 1 & 1 & -4 \end{pmatrix},$$

---

[1]Hutchinson, J. E., Fractals and self-similarity, Indiana Univ. Math. J., **30**(1981), 713–747; Hata, M., On the structure of self-similar sets, Japan J. Appl. Math., **2**(1985), 381–414.

we get $H_1 = \begin{pmatrix} T & {}^tJ \\ J & X \end{pmatrix}$, and so

$$\begin{pmatrix} (H_1 f)|_{V_0} \\ (H_1 f)|_{V_1 \setminus V_0} \end{pmatrix} = \begin{pmatrix} T & {}^tJ \\ J & X \end{pmatrix} \begin{pmatrix} f|_{V_0} \\ f|_{V_0 \setminus V_1} \end{pmatrix}.$$

The maps H_0 and H_1 are related in the following way.

LEMMA 4.1.

(4.1) $$\frac{3}{5} H_0 = T - {}^tJX^{-1}J.$$

The above formula relates the difference operator H_0 on (V_0, B_0) to the difference operator H_1 on (V_1, B_1), which we call a *renormalization equation.* It will play a key role in our effort to establish analysis on the Sierpiński gasket. One can easily verify the formula by a bread–and–butter calculation. From Lemma 4.1 one gets

LEMMA 4.2.
$$\frac{3}{5}(H_0 f)(p_i) = (H_1 f)(p_i) + \frac{2}{5} \sum_{j \neq i}(H_1 f)(q_j) + \frac{1}{5}(H_1 f)(q_j).$$

PROOF. For $f \in l(V_1)$, put $f_0 = f|_{V_0}$, $f_1 = f|_{V_1 \setminus V_0}$. Then we have

$$\frac{3}{5} H_0 f_0 = T f_0 - {}^tJX^{-1}J f_0 \qquad \text{(by Lemma 4.1)}$$

$$= (T f_0 + {}^tJ f_1) - {}^tJX^{-1}(J f_0 + X f_1)$$

$$= (H_1 f)|_{V_1} + \frac{1}{5}\begin{pmatrix} 1 & 2 & 2 \\ 2 & 1 & 2 \\ 2 & 2 & 1 \end{pmatrix}(H_1 f)|_{V_1 \setminus V_0}.$$

This completes the proof of our lemma.   □

We now go on to the definition of a *harmonic function* on the Sierpiński gasket. Classically one defines a function f to be harmonic if $\triangle f = 0$, where Δ is the Laplacian. Here we go in reverse: we first define the harmonicity of a function (Definition 4.4) and then we will show that a harmonic function f satisfies $\Delta f = 0$, where Δ is the Laplacian we are to define in §4.3.

DEFINITION 4.4. Denote by $C(K)$ the set of the continuous real-valued functions on K: $C(K) = \{f |\, f : K \to \mathbf{R},\ f \text{ is continuous}\}$. Let $f \in C(K)$. We say that the function f is harmonic if f satisfies

$$(H_m f)(p) = 0$$

for every $m \geq 1$ and every p in $V_m \setminus V_0$.

In the following theorems, we show that harmonic functions exist in sufficient abundance so that we can find a unique harmonic function satisfying any given set of boundary values (the harmonic Dirichlet problem), and that no harmonic function takes on either the maximum or the minimum value in the interior of the Sierpiński gasket (the maximum principle).

But what is the boundary of the Sierpiński gasket? If one looks at Figure 4.1, one will see that for a point p in $V_m \setminus V_0$ the number $\sharp(V_{m,p})$ of the points in $V_{m,p}$

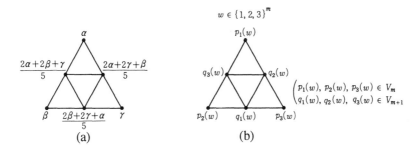

FIGURE 4.2. (a) Values of f on V_1; (b) passing from V_m to V_{m+1}

is four while $\sharp(V_{m,p})$ is two for $p \in V_0$; this makes V_0 stand out from the rest of the V_m. In this sense we shall consider V_0 as the boundary of the Sierpiński gasket.

THEOREM 4.1. *Given three numbers α, β, γ, there exists a unique harmonic function f satisfying $f(p_1) = \alpha$, $f(p_2) = \beta$ and $f(p_3) = \gamma$.*

THEOREM 4.2 (The Maximum Principle). *If a harmonic function defined on the Sierpiński gasket K attains the maximum value in the interior $K \backslash V_0$ of K, then f is constant throughout K.*

PROOF OF THEOREM 4.1. First we solve $(H_1 f)|_{V_1 \backslash V_0} = 0$ in the matrix form:

$$\begin{pmatrix} f(q_1) \\ f(q_2) \\ f(q_3) \end{pmatrix} = \frac{1}{5} \begin{pmatrix} 1 & 2 & 2 \\ 2 & 1 & 2 \\ 2 & 2 & 1 \end{pmatrix} \begin{pmatrix} f(p_1(w)) \\ f(p_2(w)) \\ f(p_3(w)) \end{pmatrix}$$

See Figure 4.2 (a). We have thus determined the values of f on V_1. Next we assume that we have the values of f on V_m. Looking at $F_w(V_0) = \{p_1(w), p_2(w), p_3(w)\}$ (Figure 4.2 (b)), we find that if f is harmonic then $(H_{m+1} f)(q_i(w)) = 0$, $i = 1, 2, 3$. Solving this in exactly the same way as we calculated the values of f on V_1 from its values on V_0, we obtain

$$\begin{pmatrix} f(q_1(w)) \\ f(q_2(w)) \\ f(q_3(w)) \end{pmatrix} = \frac{1}{5} \begin{pmatrix} 1 & 2 & 2 \\ 2 & 1 & 2 \\ 2 & 2 & 1 \end{pmatrix} \begin{pmatrix} f(p_1(w)) \\ f(p_2(w)) \\ f(p_3(w)) \end{pmatrix}.$$

With this formula we can determine inductively, starting with the values of f on V_0, the values of f on V_1, V_2, ..., ending with the map $f : V_* \to \mathbf{R}$, where $V_* = \bigcup_{m \geq 1} V_0$. We complete our proof by showing the following:

(1) f satisfies $(H_m f)(p) = 0$ for every $m \geq 1$ and every p in $V_m \backslash V_0$.
(2) We can extend f to a continuous function on K.

Proof of (1). Our proof is by induction on m. For $m = 1$, (1) holds since $(H_1 f)(q_i) = 0$, $i = 1, 2, 3$. Suppose the statement (1) is true up to m. Let $p \in V_{m+1} \backslash V_m$. Then $p = q_i(w)$ for some $w \in \{1, 2, 3\}^m$ and some i, and the construction of f implies that $(H_{m+1} f)(p) = 0$.

If $p \in V_m \backslash V_0$, we have $p \in F_w(V_0) \cap F_{w'}(V_0)$ for some $w, w' \in \{1, 2, 3\}$ with $w \neq w'$. Three situations are possible for V_m around p, which we show in Figure 4.3; they are essentially identical and so we consider only the case I.

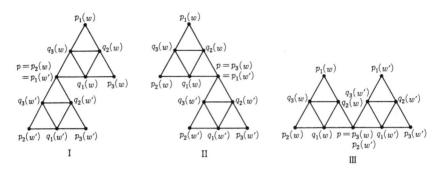

FIGURE 4.3. V_m around p

We use Lemma 4.2, substituting $p_i(w)$, $p_i(w')$ for p_i and $q_i(w)$, $q_i(w')$ for q_i, together with the fact that

$$(H_{m+1}f)(q_i(w)) = (H_{m+1}f)(q_i(w')) = 0, \qquad i = 1, 2, 3,$$

to obtain

$$\frac{5}{3}(f(p_1(w)) + f(p_3(w)) - 2f(p_2(w)))$$
$$= f(q_1(w)) + f(q_3(w)) - 2f(p_2(w)),$$

$$\frac{5}{3}(f(p_2(w')) + f(p_3(w')) - 2f(p_1(w')))$$
$$= f(q_2(w')) + f(q_3(w')) - 2f(p_1(w')).$$

Hence $(H_{m+1}f)(p) = 5/3(H_mf)(p)$, but by the induction hypothesis, $(H_mf)(p) = 0$. Therefore, $(H_{m+1}f)(p) = 0$, and the proof of (1) is complete.

Proof of (2). A simple calculation shows the following:

(i) For q_j, $j = 1, 2, 3$,

$$\min_{i=1,2,3} f(p_i) \leq f(q_j) \leq \max_{i=1,2,3} f(p_i),$$

where the equalities hold only when $f(p_1) = f(p_2) = f(p_3)$.

(ii) For $k = 1, 2, 3$,

$$\max_{1 \leq i < j \leq 3} |f(p_i(k)) - f(p_j(k))| \leq \frac{3}{5} \max_{1 \leq i < j \leq 3} |f(p_i) - f(p_j)|.$$

We use (i) and (ii) to get the following lemma by induction.

LEMMA 4.3. *Let* $w \in \{1, 2, 3\}^m$.

(i) *For* $p \in V_* \cap K_w$,

$$\min_{i=1,2,3} f(p_i(w)) \leq f(p) \leq \max_{i=1,2,3} f(p_i(w)).$$

(ii) *Set* $C = \max_{1 \leq i < j \leq 3} |f(p_i) - f(p_j)|$ *and* $v_w(f) = \max_{1 \leq i < j \leq 3} |f(p_i(w)) - f(p_j(w))|$.
Then

$$v_w(f) \leq C \left(\frac{3}{5}\right)^m.$$

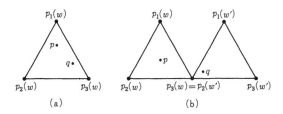

FIGURE 4.4

Select two points $p, q \in V_*$ such that $|p - q| \leq (1/2)^k$. Recall that we started the construction of the Sierpiński gasket with the equilateral triangle of unit length whose vertices were $\{p_1, p_2, p_3\}$, so that for any $w \in \{1, 2, 3\}^m$ the diameter of K_w is $(1/2)^k$. Here by the diameter of K_w we mean the maximum distance of two points in K_w. Thus when we have $|p - q| \leq (1/2)^k$, the following two possibilities arise (Figure 4.4).

(a) For some $w \in \{1, 2, 3\}$ we have $p, q \in K_w$.

(b) For some $w, w' \in \{1, 2, 3\}$ such that $K_w \cap K_{w'} \neq \emptyset$ we have

$$p \in K_w \quad \text{and} \quad q \in K_{w'}.$$

If situation (a) occurs, then Lemma 4.3 implies that $|f(p) - f(q)| \leq C(3/5)^k$. Similarly if we are in situation (b) we can show that $|f(p) - f(q)| \leq 2C(3/5)^k$. Hence we have proved

LEMMA 4.4. *If two points $p, q \in V_*$ satisfy $|p - q| \leq (1/2)^k$, then*

$$|f(p) - f(q)| \leq 2C \left(\frac{3}{5} \right)^k.$$

Now given $x \in K$, we choose a sequence $\{x_n\}_{n \geq 1}$ in V_* converging to x as $n \to +\infty$. Then for any k we can find large enough m and n such that $|x_n - x_m| \leq (1/2)^k$. So by Lemma 4.3, we have $|f(x_n) - f(x_m)| \leq 2C(3/5)^k$; hence $\{f(x_n)\}_{n \geq 1}$ is a Cauchy sequence, and hence it converges as $n \to \infty$. If x does not belong to V_* we define the value of f at x by $f(x) = \lim_{n \to \infty} f(x_n)$. This way f becomes a function on K extending the original f defined on V_*. The continuity of f is evident from the way it was constructed. This completes the proof of Theorem 4.1. $\square$

We can derive Theorem 4.2 easily from part (i) of Lemma 4.2.

4.2. The wave equation on the Sierpiński gasket. A physical observation

Keep fixed the three boundary points of the Sierpiński gasket so that the gasket lies flat. Beat the gasket from above with a drumstick to make it vibrate vertically (this "Sierpiński gasket drum" does not exist in reality). If one works with an ordinary drum, one represents the surface vibration by the wave equation,

$$M \frac{\partial^2 u}{\partial t^2} = k \left(\frac{\partial^2 u}{\partial x^2} + \frac{\partial^2 u}{\partial y^2} \right) \quad \text{on } \Omega,$$

where Ω is a bounded region in the plane corresponding to the surface of the drum, $u(x, y, t)$ denotes the position of point (x, y) at time t, M is the mass of the drum surface, and k is the coefficient of elasticity.

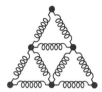 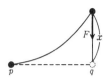

FIGURE 4.5. The model of an oscillation on (V_m, B_m)

In the above setting, $\partial^2/\partial x^2 + \partial^2/\partial y^2$ is the Laplacian on Ω. We might work out an analogy of this with the "Sierpiński gasket drum" and express its surface vibration by

$$M\frac{\partial^2 u}{\partial t^2} = k\Delta u \quad \text{on } K.$$

Here Δ is the "Laplacian" we have yet to define on the Sierpiński gasket K, and M is the mass of the Sierpiński gasket, k is the coefficient of elasticity, and $u(x, t)$ is the position of the point x in the vertical direction at time t.

Conversely, once we construct a model which describes the vibration of the Sierpiński gasket drum, we will know how we should go about defining the Laplacian on the drum surface, the Sierpiński gasket.

To carry out our plan we first consider vibration on each graph of the sequence $\{(V_m, B_m)\}$ which approximates the Sierpiński gasket, and then, passing to the natural limit, we will have vibration on the Sierpiński gasket.

(a)   **The discrete wave equation on (V_m, B_m).**

We assign a point mass $M_{m,p}$ to each vertex p of V_m. These mass points are linked by springs $(p, q) \in B_m$ with coefficients of elasticity k_m (Figure 4.5), where we assume that the vibration is vertical. Each spring deforms as shown in Figure 4.5, exerting on its neighboring mass points a force in the vertical direction. Now denote by $u(p, t)$ the *position of mass point p in the vertical direction at time t*. Then the force exerted on mass point p at time t is the combined forces of the springs connected to p, and so, if we ignore gravity, it is equal to

$$- \sum_{q \in V_{m,p}} k_m(u(p, t) - u(q, t)).$$

Hence, we may write the equation of the motion of p as

(4.2) $M_{m,p}\frac{d^2}{dt^2}u(p, t) = k_m(H_m u)(p),$

where H_m is the difference operator derived in the previous section.

We now have to discuss what sort of values of $M_{m,p}$ and $k_{m,p}$ best suit us to describe oscillations of the Sierpiński gasket, using the above model in the most natural way.

(b)   **The mass distribution $(M_{m,p})$.**

Let M be the mass of the Sierpiński gasket, which we assume to be uniformly distributed. Then for each $w \in \{1, 2, 3\}^m$ the mass of K_w is $(1/3)^m M$. For $p \in V_m$, the number of the elements w in $\{1, 2, 3\}^{m+1}$ for which $p \in K_w$ is two if $p \in V_m \backslash V_0$ and one if $p \in V_0$ (Figure 4.6).

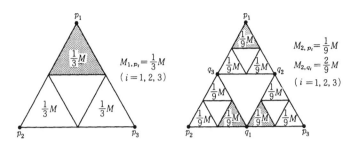

FIGURE 4.6. The mass distribution on the Sierpiński gasket

Hence, it would be quite natural to define $M_{m,p}$ by

(4.3)
$$M_{m,p} = \begin{cases} \dfrac{2}{3^{m+1}}M, & p \in V_m \backslash V_0, \\ \dfrac{1}{3^{m+1}}M, & p \in V_0. \end{cases}$$

(c)   The coefficient of elasticity k_m

For a position function $u : V_m \to \mathbf{R}$ of the points in V_m, denote by $E_m(u)$ the sum of the elastic energy stored in each spring; then we have

(4.4)
$$E_m(u) = \sum_{(p,q) \in B_m} \frac{k_m}{2}(u,(p) - u(q))^2.$$

For each $w = w_1 w_2 \cdots w_m$ in $\{1,2,3\}^m$ and each function $u : V_m \to \mathbf{R}$, let $E^w(u)$ be the following number:

(4.5)
$$E^w(u) = \sum_{i>j}(u(p_i(w)) - u(p_j(w)))^2.$$

REMARK. When $m = 0$ we think of the set $\{1,2,3\}^m$ as a singleton set whose element is $\emptyset$; $\{1,2.3\}^0 = \{\emptyset\}$. For notational convenience we make the following convention: $F_\emptyset =$ the identity map, $K_\emptyset = K$, $p_i(\emptyset) = p_i$, $q_i(\emptyset) = q_i$ $(i = 1,2,3)$. Notice in particular that we may write

$$E^\emptyset(u) = \sum_{i>j}(u(p_i)) - u(p_j))^2.$$

The definition of $E^w(u)$ implies the equation

(4.6)
$$E_m(u) = \frac{k_m}{2} \sum_{w \in \{1,2,3\}^m} E^w(u).$$

We now impose some physical assumption on this elastic energy:

Hypothesis 4.1. For every function $u : V_m \to \mathbf{R}$, $m \geq 0$, we assume that

(4.7)
$$E_m(u) = \min \big\{ E_{m+1}(u) \big| v : V_{m+1} \to \mathbf{R}, \ v|_{V_m} = u \ \big\}.$$

In this hypothesis, (4.7) gives a sufficient condition to insure the following naturally–hoped–for physical situation: For the position function $u : K \to \mathbf{R}$ on the Sierpiński gasket, the elastic energy $E(u)$ should be the limit of $E_m(u|_{V_m})$ as m

goes to $+\infty$. The value of $E(u)$ should neither diverge nor equal 0 for sufficiently many functions.

Our following discussion will establish the relation $k_{m+1} = 5/3k_m$ under Hypothesis 4.1, thus giving a natural way to define k_m.

LEMMA 4.5. (i) *For* $u : F_w(V_0) \to \mathbf{R}$ *we have*

$$E^w(u) = \frac{5}{3} \min \left\{ \sum_{j=1}^{3} E^{w_j}(v) \Big| v : F_w(V_1) \to \mathbf{R}, \; v|_{F_w(V_0)} = u \right\}.$$

The right–hand–side assumes its minimum value if and only if $(H_{m+1}v)(q_i(w)) = 0$ *for* $i = 1, 2, 3$.

(ii) *For* $u : V_m \to \mathbf{R}$ *we have*

$$\frac{1}{k_m} E_m(u) = \frac{5}{3} \frac{1}{k_{m+1}} \min \left\{ E_{m+1}(v) \Big| v : V_{m+1} \to \mathbf{R}, \; v|_{V_m} = u \right\}.$$

The right–hand–side assumes its minimum value if and only if $(H_{m+1}v)(q) = 0$ *for every* q *in* $V_{m+1} \backslash V_m$.

Before we prove Lemma 4.5 we wish to explain the physical meaning of a function v for which the value of the RHS in (ii) takes its minimum. Recall that $-k_m(H_{m+1})v(q)$ is the force on the mass point q, and so the necessary and sufficient condition that $(H_{m+1}v)(q) = 0$ hold for every $q \in V_{m+1}/V_m$ is equivalent to the fact that v restricts to a position function $u : V_m \to \mathbf{R}$ while keeping points of $V_{m+1}\backslash V_m$ in equilibrium.

PROOF. (i)   We work the case that $w = \emptyset$, $\emptyset \in \{1, 2, 3\}^0$. For $v : V_1 \to \mathbf{R}$ denote $v_0 = v|_{V_0}$ and $v_1 = v|_{V_1 \backslash V_0}$. Then we have:

$$\sum_{i=1}^{3} E^i(v) = -{}^t v H_1 v$$

$$= -({}^t v_0 \, {}^t v_1) \begin{pmatrix} T & {}^t J \\ J & X \end{pmatrix} \begin{pmatrix} v_0 \\ v_1 \end{pmatrix}$$

$$= -{}^t v_0 (T - {}^t J X^{-1} J) v_0 - {}^t (J v_0 + X v_1) X^{-1} (J v_0 + X v_1).$$

We now use (4.1) and the identity $J v_0 + X v_1 = (H_1 v)|_{V_1 \backslash V_0}$ to find that

$$\sum_{i=1}^{3} E^i(v) = -\frac{3}{5} {}^t v_0 H_0 v_0 - {}^t (H_1 v)|_{V_1 \backslash V_0} X^{-1} (H_1 v)|_{V_1 \backslash V_0}.$$

Since $-X^{-1}$ is positive definite, the minimum value of the above equation for $v_0 = u$ occurs when $(H_1 v)|_{V_1 \backslash V_0} = 0$, which is $-3/5 {}^t v_0 H_0 v_0$.

Now by the definition of $E^\emptyset$ we have

$$E^\emptyset(u) = -{}^t u H_0 u.$$

This proves (i) in case $w = \emptyset$.

If one looks at Figure 4.2 (b) one will see that the proof for a general w is exactly the same. $\square$

The proof of (ii) is evident from (i) and (4.6).

Lemma 4.5 implies immediately the following result.

THEOREM 4.3. *Hypothesis 4.1 holds if and only if, for every $m \geq 0$,*

$$k_{m+1} = \frac{5}{3} k_m.$$

With $k_0 = k$ this becomes

$$k_m = \left(\frac{5}{3}\right)^m k.$$

We have formulated the mass distribution $M_{m,p}$ and the coefficient of elasticity k_m in a most natural way for our model of the vibration of (V_m, B_m). Substituting our result in the equation of motion (4.2), we get

(4.8)

$$\text{(i)} \quad M \frac{d^2}{dt^2} u(p,t) = \frac{3}{2} \cdot 5^m k (H_m u)(p) \qquad \text{if } p \in V_m \backslash V_0,$$

$$\text{(ii)} \quad M \frac{d^2}{dt^2} u(p,t) = 3 \cdot 5^m k (H_m u)(p) \qquad \text{if } p \in V_m.$$

If we take the limit in the equation (4.8) as $m \to \infty$ we should naturally obtain the wave equation describing the vibration of the Sierpiński gasket drum, in the form

$$M \frac{\partial^2 u}{\partial t^2} = k \Delta u,$$

once we manage to define the Laplacian Δ on the Sierpiński gasket. We compare this equation with (4.8) and hope that we can define Δ by

$$\Delta u(p) = \lim_{m \to \infty} \frac{3}{2} 5^m (H_m u)(p),$$

for any point $p \in V_*$ not on the boundary V_0. This is what we justify mathematically in the next section.

4.3. The Laplacian on the Sierpiński gasket and a Gauss–Green type theorem

In the previous section we learned from a physical observation how to define the Laplacian on the Sierpiński gasket. We mathematize this definition and show that the classical Gauss-Green type formulas have their analogy on the Sierpiński gasket.

DEFINITION 4.5. For $u \in C(K)$ and $p \in V_m / V_0$, write

$$(\Delta_m u)(p) = \frac{3}{2} 5^m (H_m u)(p).$$

Suppose for some $\varphi \in C(K)$ we have

$$\max_{p \in V_m / V_0} |(\Delta_m u)(p) - \varphi(p)| \to 0 \quad \text{as } m \to +\infty.$$

Then we write $\Delta u = \varphi$ and call it the Laplacian on the Sierpiński gasket. Furthermore, we denote by $\mathcal{D}$ the set of all $u \in C(K)$ for which there exists some $\varphi \in C(K)$ such that $\Delta u = \varphi$. In other words, $\mathcal{D}$ denotes the domain of Δ.

Although we "defined" the Laplacian Δ on the Sierpiński gasket, it is not at all clear if it makes any sense. So far we do not have much information as to how many functions $\mathcal{D}$ contains, beyond the fact that any harmonic function f satisfies the equation $\Delta f = 0$, and so it is in $\mathcal{D}$. The following theorem deals with this problem.

THEOREM 4.4. *Given an arbitrary continuous function $\varphi \in C(K)$, and an arbitrary function $\rho : V_0 \to \mathbf{R}$, there exists a unique function u in $\mathcal{D}$ such that*

$$(4.9) \qquad \begin{cases} \Delta u = \varphi, \\ u|_{V_0} = \rho. \end{cases}$$

The system of equations (4.9) corresponds to the Dirichlet problem for Poisson's equation in classical analysis. We save the proof of the theorem for the next section, since it requires much background work.

DEFINITION 4.6. For $u, v : V_m \to \mathbf{R}$ define $\mathcal{E}_m(u, v)$ by

$$\mathcal{E}_m(u, v) = -\left(\frac{5}{3}\right)^m \sum_{p \in V_m} u(p)(H_m v)(p).$$

By a simple calculation we can rewrite the above equation:

$$\mathcal{E}_m(u, v) = \left(\frac{5}{3}\right)^m \sum_{(p,q) \in B_m} (u(p) - u(q))(v(p) - v(q));$$

hence, $\mathcal{E}_m(u, v)$ is a nonnegative symmetric quadratic form on $l(V_m)$, where we set $l(V_m) = \{u|\, u : V_m \to \mathbf{R}\,\}$. Recall the elastic energy we introduced in the previous section:

$$E_m = \sum_{(p,q) \in B_m} \frac{k_m}{2}(u(p) - u(q))^2.$$

Here $k_m = (5/3)^m k$, and so setting $k = 2$ we get

$$E_m(u) = \mathcal{E}_m(u, u).$$

From this with Lemma 4.5 and Theorem 4.3 we get, for $u : V_{m+1} \to \mathbf{R}$,

$$(4.10) \qquad \mathcal{E}_m(u|_{V_m}, u|_{V_m}) \le \mathcal{E}_{m+1}(u, u),$$

where the equality holds if and only if $(H_{m+1}u)(q) = 0$ for every q in $V_{m+1}\backslash V_m$.

In order to do analysis on the Sierpiński gasket one must introduce a measure and an integration with respect to it. To do this we set up the probability measure corresponding to a uniform distribution (cf. §4.2):

DEFINITION 4.7. We define a probability measure ν on K in such a way that it satisfies $\nu(K_w) = (1/3)^m$ for every $m \ge 0$ and every $w \in \{1, 2, 3\}^m$ – such a ν exists, and it is unique.

In fact, we may define ν by $\nu(A) = \mathcal{H}^d(A)/\mathcal{H}^d(K)$, $A \subset K$, where $d = \log 3/\log 2$ and $\mathcal{H}^d$ is the Hausdorff measure of dimension d.

Our next task is to define a basis for interpolation so that we can extend any function on V_m to a function on K. Note that for a given $\psi : V_m \to \mathbf{R}$ there exists a unique continuous function f on K such that $f|_{V_m} = \psi$, and $(H_n f)(q) = 0$ for every $n > m$ and every q in $V_n\backslash V_m$, or equivalently for every $w \in \{1, 2, 3\}$, $g = f \circ F_w$ is a harmonic function which satisfies $g(p_i) = \psi(p_i(w))$ for $i = 1, 2, 3$ – Theorem 4.1 insures exactly one such harmonic function. We might say that f is a continuous piecewise-harmonic extension of $\psi : V_m \to \mathbf{R}$ to K.

Under the above circumstance we make the following definition.

DEFINITION 4.8. For $m \geq 0$ and $p \in V_m$, we define ψ_p^m to be a continuous function of K such that for every $w \in \{1, 2, 3\}^m$ the composite function $\psi_p^m \circ F_w$ is harmonic on K, and for $q \in V_m$,

$$\psi_p^m(q) = \begin{cases} 1 & \text{if } q = p, \\ 0 & \text{if } q \neq p. \end{cases}$$

If f is the piecewise-harmonic extension of the function $\psi : V_m \to \mathbf{R}$ to K, Definition 4.8 evidently enables us to write

$$f = \sum_{p \in V_m} \psi(p) \psi_p^m.$$

LEMMA 4.6. (i) *For $f \in C(K)$ consider $g_m : V_m \to \mathbf{R}$, $m = 1, 2, \ldots$, such that*

$$\max_{p \in V_m} |g_m(p) - f(p)| \to 0 \quad \text{as } m \to \infty,$$

and write

$$f_m = \sum_{p \in V_m} g_m(p) \psi_p^m.$$

Then as $m \to \infty$, f_m converges uniformly to f.

(ii) *For $m \geq 0$ and $p \in V_m$ we have*

$$\int_K \psi_p^m \, d\nu = \begin{cases} 2\left(\dfrac{1}{3}\right)^{m+1} & \text{if } p \in V_m \backslash V_0, \\[2mm] \left(\dfrac{1}{3}\right)^m & \text{if } p \in V_0. \end{cases}$$

PROOF. (i). Set $\alpha_m = \max_{p \in V_m} |g_m(p) - f(p)|$. Since f is continuous on K, it is uniformly continuous there; this means by definition that there exists a sequence of numbers $\epsilon_0, \epsilon_1, \epsilon_2, \ldots$, with $\epsilon_m > 0$ and converging to 0, such that for any pair of numbers p and q we have

$$|f(p) - f(q)| \leq \epsilon_m \quad \text{if } |p - q| \leq \left(\frac{1}{2}\right)^m.$$

Now for $w \in \{1, 2, 3\}^m$ look at the difference between f_m and f on K_w. If we set $h = f \circ F_w$, $H = f_m \circ F_w$, we get

$$h(p_i) = f(p_i(w)), \quad H(p_i) = g_m(p_i(w)) \quad \text{for } i = 1, 2, 3,$$

and H is harmonic. Notice further that if $p, q \in K_w$ then $|p - q| \leq (1/2)^m$, and so for $p, q \in K$, we have $|h(p) - h(q)| \leq \epsilon_m$.

For $p \in K$, we have

$$|h(p) - H(p)| \leq |h(p) - h(p_1)| + |h(p_1) - H(p_1)| + |H(p_1) - H(p)|,$$

where we have the estimates $|h(p) - h(p_1)| \leq \epsilon_m$ and $|h(p_1) - H(p_1)| \leq \alpha_m$.

Moreover, by the maximum principle (Theorem 4.2) we have

$$\min_{i=1,2,3} H(p_i) \leq H(p) \leq \max_{i=1,2,3} H(p_i);$$

so setting $H(p_k) = \max_{i=1,2,3} H(p_i)$ and $H(p_l) = \min_{i=1,2,3} H(p_i)$, we obtain the estimate

$$\begin{aligned} |H(p) - H(p_i)| &\leq |H(p_k) - H(p_l)| \\ &\leq |H(p_k) - h(p_k)| + |h(p_k) - h(p_l)| + |h(p_l) - H(p_l)| \\ &\leq 2\alpha_m + \epsilon_m. \end{aligned}$$

From these estimates we get $|h(p) - H(p)| \leq 3\alpha_m + 2\epsilon_m$. Now we note that for any w in $\{1, 2, 3\}^m$ and any p in K_w, $|f_m(p) - f(p)| \leq 3\alpha_m + 2\epsilon_m$. Thus we have

$$\sup_{p \in K} |f_m(p) - f(p)| \leq 3\alpha_m + 2\epsilon_m \to 0, \quad \text{as } m \to \infty.$$

Hence, f_m converges uniformly to f as m tends to ∞.

(ii)   For $w \in \{1, 2, 3\}^m$ consider

$$\int_{K_w} \psi^m_{p_i(w)} \, d\nu,$$

which satisfies, by symmetry,

$$\int_{K_w} \psi^m_{p_1(w)} \, d\nu = \int_{K_w} \psi^m_{p_2(w)} \, d\nu = \int_{K_w} \psi^m_{p_3(w)} \, d\nu.$$

We also know that $\psi^m_{p_1(w)} + \psi^m_{p_2(w)} + \psi^m_{p_3(w)} \equiv 1$ on K, and hence

$$\int_{K_w} \psi_{p_i(w)} \, d\nu = \frac{1}{3} \int_{K_w} d\nu = \left(\frac{1}{3}\right)^{m+1}.$$

For $p \in V_m$, if $w \in \{1, 2, 3\}^m$ and $p \notin K_w$ we must have $\psi^m_p = 0$ on K_w. Hence,

$$\int_K \psi^m_p \, d\nu = \sum_{\substack{w \in \{1,2,3\}^m \\ p \in K_w}} \int_{K_w} \psi^m_p \, d\nu.$$

Now we go back to Figure 4.6 and count that

$$\sharp\{w \mid w \in \{1, 2, 3\}^m, \, p \in K_w\} = \begin{cases} 2 & \text{if } p \in V_m \backslash V_0, \\ 1 & \text{if } p \in V_o; \end{cases}$$

therefore, we have

$$\int_K \psi^m_p \, d\nu = \begin{cases} 2\left(\frac{1}{3}\right)^{m+1} & \text{if } p \in V_m / V_0, \\ \left(\frac{1}{3}\right)^{m+1} & \text{if } p \in V_0. \end{cases}$$

This completes the proof of (ii).   □

Our next objective is to establish a "Gauss–Green type" theorem for the Sierpiński gasket. To do this we first define the *Neumann derivative* at a boundary point $p \in V_0$ on the Sierpiński gasket, which corresponds to the derivative in the direction of the outward normal on the boundary in analysis on $\mathbf{R}^n$.

LEMMA 4.7. *Let $u \in \mathcal{D}$. Then for $p \in V_0$, $-(5/3)^m (H_m u)(p)$ converges as m tends to $+\infty$. We denote this limit by $(du)_p$, and call it the Neumann derivative of u at p.*

PROOF. It is enough to prove the lemma for $p = p_1$. We apply Lemma 4.2 to $F_{(1)^m}(V_1)$, where $(1)^m = \underbrace{11 \cdots 1}_{m \text{ times}}$, to get

(4.11) $$\frac{3}{5}(H_m u)(p_1) = (H_{m+1} u)(p_1) + \frac{2}{5} \sum_{k=2,3} (H_{m+1} u)(q^m_k) + \frac{1}{5}(H_{m+1} u)(q^m_1).$$

Here $q^m_i = q_i((1)^m)$ for $i = 1, 2, 3$.

Since $u \in \mathcal{D}$, there exists some number C such that for every integer $m \geq 1$ and every q in $V_m \backslash V_0$, the following estimate holds:

$$\left| 5^m (H_m u)(q) \right| \leq C.$$

If we multiply both sides of (4.11) by $(5/3)^{m+1}$ and use the above estimate, we get the inequality

$$\left| \left(\frac{5}{3} \right)^m (H_m u)(p_1) - \left(\frac{5}{3} \right)^{m+1} (H_{m+1} u)(p_1) \right| \leq \frac{5}{3^{m+1}} C;$$

therefore, the sequence $\{ (5/3)^m (H_m u)(p_1) \}$ is Cauchy, and so it converges as m tends to $+\infty$. $\qquad \square$

For the functions u and v in $C(K)$, denote by $\mathcal{E}(u,v)$ the limit of $\mathcal{E}_m(u,v)$, if it exists, as m tends to $+\infty$. We then have the following *Gauss–Green type theorem* for the Sierpiński gasket:

THEOREM 4.5. *For $u \in C(K)$ and $v \in \mathcal{D}$, we have*

$$\mathcal{E}(u,v) = \sum_{p \in V_0} u(p)(dv)_p - \int_K u \Delta v \, dv.$$

This theorem corresponds, for example, to the equality

$$\int_\Omega (\operatorname{grad} u, \operatorname{grad} v) \, dx \, dy = \int_{\partial \Omega} u \frac{\partial v}{\partial u} \, ds - \int_\Omega u \Delta v \, dx \, dy,$$

where Ω is a domain bounded by a smooth curve in the plane. Since we have the equality $\mathcal{E}(u,v) = \mathcal{E}(v,u)$ when u and v are in $\mathcal{D}$, we have

COROLLARY 4.6 (Gauss–Green type formulas). *Let $u, v \in \mathcal{D}$. Then*

(i) $\displaystyle \sum_{p \in V_0} (u(p)(dv)_p - v(p)(du)_p) = \int_K (u \Delta v - v \Delta u) \, dv.$

(ii) $\displaystyle \int_K \Delta u \, dv = \sum_{p \in V_0} (du)_p \qquad$ (the divergence formula).

We devote the rest of this section to the proof of Theorem 4.5.

PROOF. First we show that if $u \in C(K)$ and $v \in \mathcal{D}$, then

(4.12) $$\left(\frac{5}{3} \right)^m \sum_{p \in V_m \backslash V_0} u(p)(H_m v)(p) \to \int_K u \Delta v \, d\nu$$

as $m \to \infty$. Set

$$f_m(x) = \sum_{p \in V_m \backslash V_0} (u(p)(\Delta_m u)(p)) \psi_p^m(x) + \sum_{p \in V_0} u(p) \Delta v(p) \psi_p^m(x).$$

Then

$$\max_{p \in V_m} \left| f_m(p) - u(p) \Delta v(p) \right| \to 0 \quad \text{as } m \to \infty.$$

Hence by Lemma 4.6 (i), f_m converges uniformly to f as $m \to \infty$. Hence, by Lebesgue's theorem of bounded convergence we have

$$\int_K f_m \, d\nu \to \int_K u \Delta v \, d\nu \quad \text{as } m \to \infty.$$

By Theorem 4.6 (ii) we get

$$\int_K f_m \, d\nu = \left(\frac{5}{3}\right)^m \sum_{p \in V_m \setminus V_0} u(p)(H_m v)(p) + \sum_{p \in V_0} \left(\frac{1}{3}\right)^{m+1} u(p)\Delta v(p),$$

where

$$\sum_{p \in V_0} \left(\frac{1}{3}\right)^m u(p)\Delta v(p) \to 0$$

as $m \to \infty$. So we have verified (4.12). We now note that

$$\mathcal{E}_m(u,v) = -\left(\frac{5}{3}\right)^m \sum_{p \in V_m \setminus V_0} u(p)(H_m v)(p) + \sum_{p \in V_o} u(p)\left(-\left(\frac{5}{3}\right)^m (H_m v)(p)\right).$$

Furthermore, by Lemma 4.7,

$$\sum_{p \in V_0} u(p)\left(-\left(\frac{5}{3}\right)^m (h_m v)(p)\right) \to \sum_{p \in V_0} u(p)(dv)_p \quad \text{as } m \to \infty.$$

This and (4.12) together prove that as $m \to \infty$,

$$\mathcal{E}_m(u,v) \to \sum_{p \in V_0} u(p)(dv)_p - \int_K u\Delta v \, d\nu.$$

The proof of Theorem 4.5 is now complete. $\square$

4.4. The Dirichlet problem for Poisson's equation

We study the Dirichlet problem for Poisson's equation, which we mentioned in the previous section. Recall

THEOREM 4.4. *For a function $\varphi \in C(K)$ and a function $\rho : V_0 \to \mathbf{R}$ there exists a unique function u in $\mathcal{D}$ such that*

(4.9)
$$\begin{cases} \Delta u = \varphi, \\ u|_{V_0} = \rho. \end{cases}$$

We have

COROLLARY 4.7. *For a function $u \in C(K)$ the following two statements are equivalent.*

(i) *u is harmonic.*
(ii) *$u \in \mathcal{D}$ and $\Delta u = 0$.*

It turns out that the Dirichlet problem for Poisson's equation translated onto the Sierpiński gasket is equivalent to the infinite sequence of difference equations in the following theorem.

THEOREM 4.8. *Given functions $\varphi \in C(K)$ and $\rho : V_0 \to \mathbf{R}$, (4.9) is valid if and only if the following condition holds:*

(4.13)
$$\begin{cases} (H_m u)(p) = \left(\frac{3}{5}\right)^m \int_K \psi_p^m \varphi \, d\nu, & \text{for all } m \geq 1, \quad p \in V_m \setminus V_0, \\ \quad u|_{V_0} = \rho, \end{cases}$$

for $u \in C(K)$.

In Chapter Three we managed to express the Dirichlet problem for Poisson's equation on an interval by an infinite coefficient sequence of the equivalent difference equations ((3.19), (3.20), Theorem 3.3). The above equations show that we can extend this result to the Sierpiński gasket; that is, we can represent (not approximate) the solution of the Dirichlet problem for Poisson's equation by the solution of the difference equations obtained from discretizing its Laplacian Δ.

Now we prove Theorem 4.8.

PROOF. We first show that if (4.9) holds then (4.13) holds. Suppose that u satisfies (4.9); then Theorem 4.5 implies that for $m \geq 1$ and $p \in V_m \backslash V_0$ we have

$$(4.14) \qquad \mathcal{E}(\psi_p^m, u) = -\int_K \psi_p^m \varphi \, d\nu.$$

To continue our proof we introduce the following lemma.

LEMMA 4.8. *Suppose that the function* $\psi : K \to \mathbf{R}$ *satisfies* $(H_{m+1}\psi)(q) = 0$ *for* $q \in V_{m+1} \backslash V_m$. *Then* $\mathcal{E}_m(\psi, u) = \mathcal{E}_{m+1}(\psi, u)$, *where* $u : K \to \mathbf{R}$.

PROOF. Consider $m = 0$. Since $\psi(x) = \sum_{i=1}^3 \psi(p)\psi_{p_i}^0(x)$, it is enough to treat the special case $\psi = \psi_{p_i}^0$. We can explicitly evaluate the value of $\psi_{p_1}^0$ on V_1 by the algorithm in the proof of Theorem 4.1 (see Figure 4.2 (a)), and so, by a simple calculation, we can show that $\mathcal{E}_0(\psi, u) = \mathcal{E}_1(\psi, u)$. For a general m, we use E^w which we defined in (4.5) to obtain

$$\mathcal{E}_m(\psi, u) = \left(\frac{5}{3}\right)^m \sum_{w \in \{1,2,3\}^m} E^w(\psi, u);$$

therefore, it is enough to show that for $w \in \{1, 2, 3\}$,

$$E^w(\psi, u) = \left(\frac{5}{3}\right) \sum_{i=1}^3 E^{wi}(\psi, u),$$

but this is equivalent to the statement $\mathcal{E}_0(\psi, u) = \mathcal{E}_1(\psi, u)$ when $w = \varphi$. For a more general w, putting V_1 into correspondence with $F_w(V_1)$ will do the trick, and hence we have completed the proof of the lemma. $\qquad \square$

We now note that ψ_p^m satisfies $(H_n\psi_p^m)(q) = 0$ for $n > m$ and $q \in V_n \backslash V_m$. By Lemma 4.8, we have

$$\mathcal{E}_m(\psi, u) = \mathcal{E}_{m+1}(\psi, u) = \mathcal{E}_{m+2}(\psi, u) = \cdots;$$

hence, we get

$$\mathcal{E}(\psi, u) = \mathcal{E}_m(\psi, u) = -\left(\frac{5}{3}\right)^m \sum_{q \in V^m} \psi_p^m(q)(H_m u)(q)$$

$$= -\left(\frac{5}{3}\right)^m (H_m u)(p).$$

We substitute this in (4.14) to show that (4.9) implies (4.13).

We next show that (4.13) implies (4.9).

Note that φ is continuous in K, and so it is uniformly continuous. In this case, as in the proof of Corollary 4.6 (i), there exists a sequence of numbers $\epsilon_1, \epsilon_2, \ldots,$

with $\epsilon_i > 0$ and $\lim_{n\to\infty} \epsilon_n = 0$, such that for $p, q \in K$,

$$|p - q| \le \left(\frac{1}{2}\right)^m \Rightarrow |\varphi(p) - \varphi(q)| \le \epsilon_m.$$

Now for $p \in V_m \setminus V_0$ there are two w's such that $p \in K_w$, which we denote by w^1 and w^2. Then

$$\{\, x : \psi_p^m(x) > 0 \,\} \subseteq K_{w^1} \cup K_{w^2}.$$

Furthermore, if $x \in K_{w^1} \cup K_{w^2}$ then $|x - p| \le (1/2)^m$, and so $|\varphi(p) - \varphi(x)| \le \epsilon_m$. Thus, it follows from Corollary 4.6 (ii) that

$$\left| \int_K \psi_p^m \varphi(x)\, d\nu(x) - \int_K \psi_p^m(x)\varphi(p)\, d\nu(x) \right|$$

$$\le \epsilon_m \int_K \psi_p^m(x)\, d\nu(x) = 2\epsilon_m \left(\frac{1}{3}\right)^{m+1};$$

hence, we have

$$|(\Delta_m u)(p) - \varphi(p)| = \frac{3^{m+1}}{2} \left| \int_K \psi_p^m \varphi\, d\nu - 2\left(\frac{1}{3}\right)^{m+1} \varphi(p) \right| \le \epsilon_m,$$

which shows that

$$\max_{p \in V_m \setminus V_0} |(\Delta_m u)(p) - \varphi(p)| \to 0 \quad \text{as } m \to \infty.$$

This completes the proof of Theorem 4.8. $\qquad\qquad\qquad\qquad\qquad\square$

By virtue of Theorem 4.8 we may compute the values of the solution u on $V_* = \bigcup_{m \ge 0} V_m$ of Dirichlet's problem for the Poisson equation on the Sierpieński gasket by the following algorithm:

(i) Set $u|_{V_0} = \rho$.

(ii) Once we know the values of u on V_m, the following holds for each $w \in \{1, 2, 3\}^m$:

$$(4.15) \qquad (H_{m+1} u)(q_i(w)) = \left(\frac{3}{5}\right)^{m+1} \int_K \psi_{q_i(w)}^{m+1} \varphi\, d\nu.$$

This is a system of equations with respect to $u(q_i(w))$ $(i = 1, 2, 3)$, and so when the value of u at each $p_i(w) \in V_m$ is given we can determine $u(q_i(w))$; therefore, we have the values of u on V_{m+1}.

The proof of Theorem 4.4 reduces, therefore, to showing the following:

(1) The function $u : V_* \to \mathbf{R}$ given by the above algorithm extends uniquely to a continuous function on K.

(2) The function $u : V_* \to \mathbf{R}$ given by the above algorithm satisfies (4.13) for every $m \ge 1$ and every $p \in V_m \setminus V_0$.

Proof of (1): As the function φ is continuous on K, it is bounded there. We also have from Lemma 4.6 (ii) that

$$\int_K \psi_p^{m+1}\, d\nu = 2\left(\frac{1}{3}\right)^{m+2}, \qquad p \in V_{m+1} \setminus V_0;$$

hence, the right-hand side of (4.15) satisfies

$$\left| \left(\frac{3}{5}\right)^{m+1} \int_K \psi_{q_i(w)}^{m+1} \varphi\, d\nu \right| \le \left(\frac{1}{5}\right)^m C,$$

where C is some constant independent of m, w, and $q_i(w)$. Thus, if for $q \in V_{m+1} \backslash V_m$ we set

$$\left(\frac{3}{5}\right)^{m+1} \int_K \psi_q^{m+1} \varphi \, d\nu = \left(\frac{1}{5}\right)^m C(q),$$

we get $|C(q)| \leq C$.

Write

$$u_m = \sum_{p \in V_m} u(p) \psi_p^m,$$

where u is the function defined by the algorithm above. Rewrite (4.15) as

$$J u_0^w + X u_1^w = \left(\frac{1}{5}\right)^m C_w,$$

where

$$u_0^w = \begin{pmatrix} u(p_1(w)) \\ u(p_2(w)) \\ u(p_3(w)) \end{pmatrix}, \quad u_1^w = \begin{pmatrix} u(q_1(w)) \\ u(q_2(w)) \\ u(q_3(w)) \end{pmatrix}, \quad C_w = \left(\frac{1}{5}\right)^m \begin{pmatrix} C(q_1(w)) \\ C(q_2(w)) \\ C(q_3(w)) \end{pmatrix}.$$

We then have

$$u_1^w = -X^{-1} J u_0^w + X^{-1} \left(\frac{1}{5}\right)^m C_w.$$

Set $(-X^{-1} J u_0^w)_{q_i(w)} = \bar{u}_m(q_i(w))$; this gives the value of u on V_m which is the solution of $(H_m \bar{u})(q_i(w)) = 0$, and hence we get

$$\bar{u}_m(q_i(w)) = u_m(q_i(w)).$$

It follows that

$$u_{m+1} = \sum_{p \in V_{m+1}} u_m(p) \psi_p^{m+1} + \sum_{p \in V_{m+1} \backslash V_0} v_m(q),$$

where we set

$$\left(\left(\frac{1}{5}\right)^m X^{-1} C_w\right)(q_i(w)) = v_m(q_i(w)).$$

Here $\sum_{p \in V_{m+1}} u_m(p) \psi_p^{m+1} = u_m$, and so setting $v_m = \sum_{p \in V_{m+1} \backslash V_0} v_m(q)$, we get

$$u_{m+1} = u_m + v_m.$$

Thus, we obtain

$$\sup_{x \in K} |u_{m+1}(x) - u_m(x)| \leq \sup_{x \in K} |v_m(x)| \leq \left(\frac{1}{5}\right)^m C.$$

Since u_m converges uniformly as $m \to \infty$, the function $u : V_* \to \mathbf{R}$ extends uniquely to a continuous function on K.

Proof of (2): We induct on m. For $m = 1$, we have $V_1 \backslash V_0 = \{q_1, q_2, q_3\}$; so setting $w = \emptyset$ in (4.15), we evidently establish (4.13) for $V_1 \backslash V_0$. Suppose that (4.13) holds for all natural numbers up to m. For $q \in V_{m+1} \backslash V_m$, (4.15) is just (4.13). The difficulty arises for the case $p \in V_m \backslash V_0$. Suppose that $p \in F_w(V_1) \cap F_{w'}$. Then the arrangement of V_{m+1} about p is as in Figure 4.3 (the proof of Theorem 4.1, §4.1). We investigate case I of Figure 4.3. We apply Lemma 4.2 to $F_w(V_1)$ and $F_{w'}$ to see that

$$\frac{3}{5}(H_m u)(p) = (H_{m+1} u)(p) + \frac{2}{5} \sum_{q \in Q_1} (H_{m+1} u)(q) + \frac{1}{5} \sum_{q \in Q_2} (H_{m+1} u)(q),$$

where $Q_1 = \{q_3(w), q_1(w), q_2(w'), q_3(w')\}$, $Q_2 = \{q_2(w), q_3(w')\}$.

Our induction hypothesis, coupled with the fact that (4.15) holds for $q \in V_{m+1} \backslash V_m$, gives the equality

$$\left(\frac{3}{5}\right)^{m+1} \int_K \psi_p^m \varphi \, d\nu = (H_{m+1}u)(p) + \frac{2}{5} \sum_{q \in Q_1} \left(\frac{3}{5}\right)^{m+1} \int_K \psi_q^{m+1} \varphi \, d\nu$$

$$+ \frac{1}{5} \sum_{q \in Q_2} \left(\frac{3}{5}\right)^{m+1} \int_K \psi_q^{m+1} \varphi \, d\nu.$$

Hence, we have

$$(H_{m+1}u)(p) = -\left(\frac{3}{5}\right)^{m+1} \int_K \psi_p^{m+1} \varphi \, d\nu,$$

which shows that (4.13) holds for $p \in V_m \backslash V_0$. This completes the proof of (2).

The proof of Theorem 4.4 follows from the proof of (1) and (2).

Exercises

4.1. Determine the Hausdorff dimension of the Sierpiński gasket (Hint: find a suitable open set V and show that it satisfies the open–set condition (§2.2, Chapter 2).

4.2. (1) Let

$$A_1 = \begin{pmatrix} 1 & 0 & 0 \\ \frac{1}{5} & \frac{2}{5} & \frac{1}{5} \\ \frac{1}{5} & \frac{1}{5} & \frac{2}{5} \end{pmatrix}, \quad A_2 = \begin{pmatrix} \frac{2}{5} & \frac{1}{5} & \frac{1}{5} \\ 0 & 1 & 0 \\ \frac{1}{5} & \frac{1}{5} & \frac{2}{5} \end{pmatrix}, \quad A_3 = \begin{pmatrix} \frac{2}{5} & \frac{1}{5} & \frac{1}{5} \\ \frac{1}{5} & \frac{2}{5} & \frac{1}{5} \\ 0 & 0 & 1 \end{pmatrix},$$

and show that a harmonic function f on the Sierpiński gasket satisfies the equality

$$\begin{pmatrix} f(p_1(w)) \\ f(p_2(w)) \\ f(p_3(w)) \end{pmatrix} = A_{w_m} A_{w_{m-1}} \cdots A_2 A_1 \begin{pmatrix} f(p_1) \\ f(p_2) \\ f(p_3) \end{pmatrix},$$

where $w = w_1 w_2 \cdots w_m \in \{1, 2, 3\}^m$.

(2) Consider a harmonic function f on the Sierpiński gasket satisfying $f(p_1) = 1$, $f(p_2) = f(p_3) = 0$. For $k = 1, 2, 3$ define $(k)^m \in \{1, 2, 3\}^m$ by $(k)^m = \underbrace{kk \cdots k}_{m}$.

Show that

$$f(p_2((1)^m)) = f(p_3((1)^m)) = 1 - \left(\frac{3}{5}\right)^m,$$

$$f(p_1((2)^m)) = \frac{1}{2}\left(\frac{3^m + 1}{5^m}\right), \qquad f(p_3((2)^m)) = \frac{1}{2}\left(\frac{3^m - 1}{5^m}\right).$$

4.3. Let u be a harmonic function on the Sierpiński gasket. Show that for $p \in V_*$ there exists a certain $C_p > 0$ such that the following inequality holds for $q \in K$:

$$|u(p) - u(q)| \leq C_p |p - q|^\alpha,$$

where $\alpha = \log(5/3)/\log 2$.

4.4. Suppose that u is the restriction of a C^2–function on $\mathbf{R}^2$ to the Sierpiński gasket. Show that $u \in \mathcal{D}$ only if u is constant on the Sierpiński gasket.

4.5. Let u be a harmonic function on the Sierpiński gasket and let p be an element of V_0.

(1) Show that $(H_{m+1}u)(p) = (3/5)(H_m u)(p)$ (Hint: Cf. the proof of Lemma 4.7).

(2) Show that $(du)_p = -(H_0 u)(p)$.

4.6. Using (4.13), derive the values of the solution u on V_0 of the Dirichlet problem for Poisson's equation

$$\begin{cases} \Delta u = 1, \\ u|_{V_0} = 0. \end{cases}$$

4.7. Suppose that the function $v : V_* \to \mathbf{R}$ is bounded on V_*. Show that for $0 < \alpha < 1$ the infinite sequence of difference equations

$$\begin{cases} u|_{V_0} = v|_{V_0}, \\ (H_m u)(p) = \alpha^m v(p), \qquad p \in V_m \backslash V_{m-1}, \end{cases}$$

has a unique solution u on K (Hint: Cf. the proof of Theorem 4.4 (1)).

4.8. The Neumann problem for the Poisson equation on the Sierpiński gasket: Given $\tau : V_0 \to \mathbf{R}$ and $\varphi \in C(K)$, the system of equations

$$\begin{cases} \Delta u = \varphi, \\ (du)_{p_1} = \tau(p_i), \qquad i = 1, 2, 3, \end{cases}$$

has a solution if and only if the equality,

$$\int_K \varphi \, d\nu = \sum_{i=1}^3 \tau(p_i),$$

holds (Hint: Use Corollary 4.6 (ii)).

Recommended Reading

[1]   Yamaguti, M., *An Introduction to Chaos and Fractals*, a text for the University of the Air, The Society for the Promotion of the University of the Air, 1992.

When we wrote *Mathematics of Fractals* in Japanese we had this elementary textbook in mind to complement the first three chapters. Unfortunately the book has not been translated into English, and we cannot think of anything written in English comparable to it.

Chapters One and Two.

[2]   Falconer, K. J., *The Geometry of Fractal Sets*, Cambridge Univ. Press, 1985.

This is the best book with which to pursue further study of the Hausdorff measure in geometric measure theory. It has a compact survey of classical results as well, together with an extensive bibliography.

[3]   Falconer, K. J., *Fractal Geometry – Mathematical Foundations and Applications*, Wiley, 1990.

Falconer presents various results concerning non-integer dimensions, at the same time putting an emphasis on applications. The book has a rich collection of topics in dynamical systems, including complex cases and random fractals.

[4]   *Fractal Geometry and Analysis*, (Edited by Bélair, J. and Dubuc, S.), Proc. of NATO ASI Series C Vol. 346, Kluwer Academic Publishers, 1991.

This proceedings (of a meeting held in Montreal in 1989) contains the invited talks, including the most recent results on subjects in measure theory, dimension theory, structures of fractals, complex dynamical systems, multi-fractals, *etc.*

Chapter Three.

[5]   Nishida, T., Mimura, M., Fujii, H., eds., *Patterns and Waves, Studies in Mathematics and its Application*, Kinokuniya & North-Holland, 1986: Uchiki, S., Chaotic Phenomena and Fractal Object, 221–258. Hata, M., Fractals in Mathematics, 259–278.

[6]   Hata, M., Yamaguti, M., Weierstrass's function and chaos, Hokkaido J. Math., 12(1983), 333–343.

[7]   Meyer, Y., translated by Ryan, R. D., *Wavelets: Algorithms and Applications*, SIAM, 1993.

[8]   Sekiguchi, T., Shiota, Y., Hausdorff dimension of graphs of some Rademacher series, Japan J. Appl. Math., 7(1990), 121–129.

[9]   Sekiguchi, T., Shiota, Y., A generalization of Hata–Yamaguti's results on the Takagi function, Japan J. Indust. Appl. Math., 8(1991), 203–219.

Chapter Four.

The fractal analysis we studied in Chapter Four is still at its early developing stage and there is no comprehensive reference in the literature (to the authors' knowledge). In the following the reader will find discussion of the current research activities and their future development in this field.

[10]   Kigami, J., Laplacians on self-similar sets – Analysis on fractals, Amer. Math. Soc. Transl. (2), Vol. 161(1994), 75–93.

The reader wishing to study the subject further might consult the references in [10].

[11]   Kigami, J., Laplacians on self-similar sets and their spectral distributions, in *Fractal Geometry and Stochastics*, eds. Bandt *et al.*, **Progress in Probability 37**, Birkhauser, 1995, 221–238.

There are many other interesting articles in *Fractal Geometry and Stochastics*.

Index

Selected Titles in This Series

(*Continued from the front of this publication*)

(See the AMS catalog for earlier titles)